Biology at Work

THE RUTGERS SERIES IN HUMAN EVOLUTION

Kingsley R. Browne, *Biology at Work: Rethinking Sexual Equality*

David Haig, *Genomic Imprinting and Kinship*

John F. Hoffecker, *Desolate Landscapes: Ice-Age Settlement in Eastern Europe*

John T. Manning, *Digit Ratio: A Pointer to Fertility, Behavior, and Health*

Paul Rubin, *Darwinian Politics*

Biology at Work
Rethinking Sexual Equality

Kingsley R. Browne

Rutgers University Press
New Brunswick, New Jersey, and London

Library of Congress Cataloging-in-Publication Data
Browne, Kingsley.
Biology at work : rethinking sexual equality / Kingsley R. Browne.
p. cm. — (The Rutgers series in human evolution)
Includes bibliographical references and index.
ISBN 0-8135-3053-9 (alk. paper)
1. Sex differences (Psychology) 2. Sexual division of labor. 3. Sex differences. I. Title. II. Series.

BF692.2 .B77 2002
305.3—dc21 2001048839

British Cataloging-in-Publication information is available from the British Library.

Manufactured in the United States of America

To Cecily and Veronica

Contents

Acknowledgments

Every scholar must acknowledge that his contribution to the literature, however modest, is necessarily dependent upon the work of many others. Especially is this so in a work such as this, which does not convey any original research findings but rather seeks to synthesize findings of others from disparate disciplines. For that reason, I would like to acknowledge all of those who produced the research upon which I rely.

More specifically, I would like to express my appreciation to those who have provided me with various kinds of personal assistance and support. My inadequate memory and abysmal record-keeping skills guarantee that the list will be incomplete. Much of my debt is owed to the hundreds of people with whom I have discussed the subjects covered by this book over the course of years.

Many people, most of whom I have never met, were kind enough to provide me with copies of papers or answer questions about their work or the work of others. These include Sheri Berenbaum, Suzanne Bianchi, Devon Brewer, Rogers Elliott, Mark Flinn, David Geary, Madeline Heilman, Michael Mills, Solomon Polachek, and Irwin Silverman. They bear no responsibility, of course, for the use to which I put their work.

Others were kind enough to provide comments on drafts of some of my earlier work that served as the basis for portions of this book. These include David Buss, John Dolan, Joseph Grano, Leroy Lamborn, Ariel Levi, Bobbi Low, Ralph Slovenko, and Lionel Tiger.

Much appreciation is also due Margaret Gruter of the Gruter Institute for Law and Behavioral Research not only for her support of scholarship on the intersection of law and biology but also for her nurturing of a community of scholars.

Special thanks must go to several people upon whom I have imposed repeatedly for counsel and/or who have suffered through drafts of this manuscript. These include Christina Beaton, Stephen Colarelli, Helena Cronin,

Oliver Curry, Patricia Hausman, Owen Jones, and Michael McIntyre. Their generosity with their time has been a resource of inestimable value.

Finally, my heartfelt thanks must go to my wife, Cynthia, and my daughters, Cecily and Veronica, who supported this research in more ways than can be counted.

Biology at Work

1 Introduction

Despite a virtual orthodoxy in the social sciences for much of the last century, according to which observed differences between the sexes are mere "social constructions," evidence from evolutionary biology and psychology is coalescing to provide a very different picture. The human brain, like the human body (and also like all other mammalian brains), is sexually dimorphic as a consequence of selection pressures experienced by ancestral generations. The sexual division of labor, a cultural universal, appears to be at least in part a consequence of this sexual dimorphism.

The formal division of labor is breaking down in Western societies, with women increasingly penetrating many formerly all-male preserves. Nonetheless striking disparities remain. Women are sparsely represented at the highest level of corporate hierarchies; many jobs continue to be largely sex segregated; and female employees earn, on average, less than men. The variegated pattern of female progress suggests causes more complex than such commonly blamed systemic factors as patriarchy and sexism. Recognition of biological contributors to contemporary workplace patterns does not establish that the patterns are either desirable or immutable, but an understanding of their origins may inform our judgments about whether intervention is appropriate and, if so, what form that intervention might take.

In the 1985 movie based upon H. Rider Haggard's *King Solomon's Mines*, the hero, Allen Quatermain, encounters an African tribe that has adopted the unusual cultural practice of living upside down.[1] They court upside down; they fight upside down; they even do laundry upside down. "Unhappy with the world the way it is," we are told, "they live upside down hoping to change it." Despite their unusual mode of life, they seem an eminently happy and well-adjusted lot.

Might there be—perhaps in still-remote parts of Africa or New Guinea—a group of real humans that has chosen to live in such a way? Despite the dizzying array of cultural practices chronicled by ethnographers, ranging from the charming to the bizarre, we can be quite certain that the answer is no. Our confidence does not depend upon the fact that so much of the planet has already been canvassed by anthropologists and others; we would be virtually as certain if there were still vast inhabited continents that had never been visited by Western scientists or explorers. That certainty would derive from our understanding of what it means to be human. As philosopher David Hume commented, we would immediately denounce as a liar a traveler returning from a far country with tales of "men, who were entirely divested of avarice, ambition, or revenge [and] who knew no pleasure but friendship, generosity, and public spirit" because such a claim would be no more plausible than "stories of centaurs and dragons, miracles and prodigies."[2] A tale of a people who lived upside down should be no more eagerly accepted. Human anatomy, physiology, and psychology are all oriented toward a "right-side up" existence, so that we need not specify for any given culture whether "right-side up" means "head up" or "feet up." By describing them as human, we have already specified which end is up.

What about a different kind of society? What about a society in which women have a monopoly on political power; where women out-compete men in the quest for positions of high status; where women leave their babies with their husbands who stay home and keep house; where the army, navy, air force, and marines are made up mostly of women; and where almost all mathematicians, physicists, and engineers are women? Is such a society any more likely than the one found by Allen Quatermain? For those who believe that sex roles are arbitrary products of social conditioning, current cultural patterns could be reversed to provide mirror images of familiar patterns. For those who believe that the human mind is a legacy of our evolutionary history as primates and mammals, however, a completely sex-role–reversed society is no more likely than one that chooses to live upside down.

In the early 1930s, anthropologist Margaret Mead returned from New Guinea to a credulous world with tales of a preliterate equivalent of the sex-role reversal described above. In her influential book *Sex and Temperament in Three Primitive Societies,* she recounted her experience with three tribes: the Arapesh, in which both males and females exhibited a "feminine" gentleness; the Mundugumor, in which both males and females exhibited a "masculine" ruthlessness and aggressivity; and the Tchambuli (or "Chambri"), who, according to Mead, showed "a genuine reversal of the sex-attitudes of our own culture, with the woman the dominant, impersonal, managing partner, the man the less responsible and the emotionally dependent person."[3] Never mind that her story had the feel of "Goldilocks and the Three Bears"—she set out to determine if the stereotypes about males and females adopted by all societies then known would hold true in three randomly selected societies and she found one society that was all "feminine," one that was all "masculine," and one that was "just right" (role-reversed). Never mind also that her recorded observations did not support her sweeping conclusions.[4]

Like her earlier account of Samoa, in which she suggested that adolescence need not be a period of *Sturm und Drang* but instead can be a conflict-free phase of liberal sexual experimentation for both sexes,[5] Mead's conclusion that "human nature is almost unbelievably malleable, responding accurately and contrastingly to contrasting cultural conditions" fell on fertile ground.[6] For much of the twentieth century, social science operated under the influence of the view of Mead and others that human nature is a *tabula rasa* upon which society can write at will.

This model of human nature and human behavior, which has been labeled the "Standard Social Sciences Model" ("SSSM"), attributes human behavior largely, if not exclusively, to social conditioning. Social phenomena are thus to a large extent divorced from the individual constituents of society—human beings motivated by a distinctive human psychology.

The Standard Social Sciences Model and Its Challengers

The SSSM elevates humans to a position unique in the animal kingdom. Although adherents of that model would readily attribute an inherent "nature" to all other animals, they are reluctant to do so for humans. They surely would not expect a hyena raised in a group of chimpanzees to attempt to walk on its knuckles, swing from the trees, or mate with a chimpanzee. But when it comes to humans they still, at least implicitly, embrace the extreme behaviorism of John Watson, who asserted "that there is no such thing as inheritance of capacity, talent, temperament, mental constitution and characteristics" but rather that these traits "depend on training that goes on mainly in the cradle."[7]

The SSSM has perhaps been most influential in governing discourse about the sexes. Just as the model denies any important role for a "human nature" that may channel human behavior and social institutions, it even more emphatically denies a role to a dual human nature—that is, a "male nature" and a "female nature." Notions of a distinctive male and female nature are simply "social constructs" built by society (at least by the half of society that is male) on the fragile foundation of undeniable and irreducible physiological differences relating to reproduction. Apart from reproductive functions, the observed differences between men and women are seen as products of different social forces acting upon males and females. For some, even recognition of two sexes is an arbitrary cultural convention. Biologist Anne Fausto-Sterling, for example, has asserted that "Western culture is deeply committed to the idea that there are only two sexes"[8] and sociologist Judith Lorber contends that "in Western societies, we see two discrete sexes and two distinguishable genders because our society is built on two classes of people, 'women' and 'men.'"[9] Neither writer, however, identifies any culture that recognizes some other number.

Notwithstanding assertions to the contrary, human societies exhibit certain predictable regularities that derive from our common nature.[10] While some patterns—such as living right-side up—are based upon a common human anatomy or physiology, many of the more interesting ones are based upon a common human psychology. Some patterns are universal, such as the sexual division of labor and systems of marriage, kinship, and religion. Some are nearly so, such as a marriage system that limits women to a single husband at a time. Others might more properly be described as "tendencies" or "predispositions," such as marriage systems that permit men to have more than one wife, a pattern that prevails in a substantial majority of cultures known to ethnographers but that governs the lives of a much smaller proportion of human beings because of the state-imposed monogamy that exists in most of the largest societies. Anthropological studies focusing on cultural regularities, rather than on the idiosyncrasies that anthropologist Robin Fox has called "ethnographic dazzle," suggest that the *tabula* may not be so *rasa* after all.

Modern evolutionary biology and psychology pose an even more direct challenge to the SSSM, with their insight that human behavioral predispositions are ultimately attributable to the same cause as the behavioral predispositions of other animals—evolution through natural selection. The centrality of mating and reproduction to evolutionary success, coupled with the differential investment of mammalian males and females in offspring, makes behavioral and temperamental identity of the sexes highly improbable. Just as no farmer expects to see identical patterns of behavior from the mare as from the stallion, from the cow as from the bull, or from the hen as from the rooster, no social scientist should expect to see identical patterns of behavior from men and women.

Claims for the existence of a recognizable "human nature" or for predictable behavioral differences between the sexes should be inherently suspect only to those who believe that the forces that created humans were importantly different from those that created the rest of the animal kingdom. If males and females are at their core psychologically identical, they are unique among mammals. This is not to deny the importance of social influences or the fact that societies have certain emergent characteristics that no amount of atomistic study of individuals could ever predict. But it is critical to understand that some social practices are more likely to arise than others precisely because human *minds* are more likely to settle on some social practices than others and that males and females tend to have different psychologies independent of the influence of cultures that expect them to be different.

The Division of Labor by Sex

A proper understanding of psychological sex differences would go far toward an understanding of the modern workplace, the study of which has heretofore been heavily biased toward the SSSM orientation. One human universal that is apparently a product of human nature is the division of labor by sex. All societies label some work "men's work" and other work "women's work." Although the content of the categories is by no means fixed—what some cultures label "men's work" is "women's work" in others—there are, nonetheless, some consistent patterns. Big-game hunting and metalworking are almost always "men's work" and cooking and grinding grain are almost always "women's work." While some divisions are obviously related to physical capacity, this is not always the case. For example, carrying water is almost always "women's work," and manufacture of musical instruments is almost always "men's work."[11]

Modern Western societies are breaking down these age-old divisions, so that workers increasingly find themselves in what anthropologists call an "evolutionarily novel environment"—an environment that differs from that in which our hominid ancestors evolved—in this case a workplace environment in which men and women work side by side and compete for position in the same status hierarchies. Today, almost all positions in the labor market are formally open to women, the primary exception being certain combat positions in the military.[12] Nonetheless, a high degree of de facto occupational segregation continues to exist, so that in practice there are many occupations that remain "men's work" and "women's work." Thus, most men work mostly with other men, and most women work mostly with other women. Moreover, even in largely integrated occupations, men are more likely than women to achieve the highest organizational positions.

The architects of sexual equality appear to have assumed that lifting formal barriers to women in the workplace would result in parity with men because men and women inherently have identical desires and capacities. When prohibitions on formal discrimination have not resulted in sexual parity, hidden discrimination is often assumed responsible. If hidden discrimination can be disproved, then informal barriers, such as sexist attitudes of parents or teachers are identified as the culprit. If direct external forces must finally (and reluctantly) be abandoned because the paths that women's lives have taken must be attributed to their own choices, then their choice becomes a "choice" that is attributed to their internalization of "patriarchal" notions about the proper role of the sexes and to their life constraints. While the causal attribution may shift over time, what does not change is the persistent invocation of causes other than women's inherent predispositions. Given the human propensity for self-deception, it may not be possible to answer the question whether these shifting arguments reflect actual beliefs or are merely opportunistic arguments to advance a political agenda.

The social-role view of sex differences is that "men and women have inherited essentially the same evolved psychological dispositions" and that behavioral sex differences are simply results of "two organizing principles of human societies: the division of labor according to sex and gender hierarchy."[13] How is it that a sexually monomorphic mind came up with the division of labor by sex and gender hierarchy? Certainly the social explanation is not the most parsimonious explanation for sex differences in behavior. Humans evolved from other creatures surely having sexually dimorphic minds. The notion that humans evolved away from the primate pattern of behavioral sex differences—presumably because it was advantageous to do so—but simultaneously replaced the preexisting biological pattern with cultural patterns having the same effect is difficult to credit. Moreover, the direction of causation in this explanation is implausibly unidirectional. Even if behavioral sex differences originated from a sexually monomorphic mind, one would expect that they would be reinforced through selection over the hundreds of thousands or millions of years that these social phenomena existed.

Stasis and Change

Trends in women's work-force participation are not easily explained in terms of broad themes such as "patriarchy," "subjugation of women," or even the waning power of a monolithic male hierarchy. The progress of women has not been uniformly slow or uniformly fast, as might be expected if it were solely a consequence of such wide-ranging forces; instead, the pattern has been much more complex, and it is that pattern that any theory of workplace sex differences must attempt to explain.

In some respects, the role of women in the work force has been massively transformed in just a few decades. In 1960, women constituted just one-third of the American work force compared to over 46 percent today. During that same period, the percentage of married women who work doubled to 61 percent. Only 4 percent of lawyers in 1970 were women, while today the figure for law school graduates exceeds 42 percent. The percentage of female physicians increased from 10 to 24 percent between 1970 and 1995, and the percentage of female medical students now exceeds 40 percent. In business, the change has been no less impressive. In 1972, women held only 18 percent of managerial and administrative positions, compared to 43 percent of such positions in 1995. These changes represent a genuine revolution in the American workplace.

Despite these striking advances, however, women are far from achieving parity in a number of areas. They constitute only 5 to 7 percent of senior executives in the largest corporations, and the average full-time female employee makes less than 75 cents for every dollar earned by the average full-time male, if factors that influence wages such as hours worked and nature of the occupation are not considered. Many occupations remain highly sex segregated. Among the occupations in the United States that remain 90 percent or more female are bank teller, receptionist, registered nurse, and preschool and kindergarten teacher. Among the occupations that are less than 10 percent female are engineer, firefighter, mechanic, and pest exterminator. Large numbers of women pursue education in some scientific fields—such as biology and medicine—yet far fewer are found in other scientific fields—such as mathematics, physics, and engineering. Despite frequent assertions that women are victims of widespread discrimination, for the past two decades unemployment rates of the two sexes have not diverged by as much as a percentage point.[14] Thus, women's progress has not been uniformly stifled nor has it uniformly advanced; instead it has been quite patchy.

The question is why. Part of the answer lies in the sexually dimorphic human mind. The disappointment of the "integrationists"—those who believe that parity of representation of men and women in all positions is a goal both admirable and achievable—is a predictable consequence of their failure to appreciate the full contours of human nature. For reasons having substantial roots in evolutionary biology, men and women differ in fundamental ways (beyond mere differences in physical strength and reproductive capacity). These differences have significant workplace implications, as average sex differences in temperament, cognitive capacities, values, and interests are causally related to average sex differences in workplace outcomes.

Temperamental differences between the sexes may result in differences in the vigor with which men and women compete for high positions. Well-known stereotypes of men as more competitive, more driven toward seeking status and resources, and more inclined to take risks than women, and stereotypes of women as more nurturing, risk averse, and less single-minded than men are true as generalizations and have an underlying biological basis. Those individuals, whether male or female, who are inclined toward competition, risk taking, and status seeking are more likely to reach the pinnacle of organizational hierarchies than those who are not.[15]

Average cognitive differences between the sexes have an impact on their average suitability for particular jobs. Males tend to outperform females in mathematical reasoning, science, mechanical comprehension, and many types of spatial ability, while females tend to outperform males in such areas as language use, reading comprehension, spelling, and mathematical computation. Because different occupations place different demands on their incumbents, we should not be surprised if groups differing in average abilities tend to gravitate toward different jobs.

Even beyond differences in skills and abilities, sex differences in interests and values may influence what kinds of jobs men and women prefer. Men, more than women, prefer to work with "things," while women, more than men, prefer to work with people, for example. Just as different occupations attract people with different abilities, they also tend to attract people with different patterns of values, once again leading to an average difference in occupational preferences.

Sex differences in sexual attitudes and behavior also have workplace implications. Stereotypes of men as being more interested in casual sexual relationships and more willing to use coercion to achieve them are not just creations of Hollywood and Madison Avenue. The sexual conflict that often occurs when men and women interact in the workplace is a predictable, if regrettable, consequence of a conflict between male and female sexual psychologies, and it is often that conflict—rather than an ideology of patriarchy—that manifests itself as sexual harassment.

A major caution is in order. In most places in this book, nothing is implied about "all women" or "all men." The two sexes overlap substantially in the traits described; the differences are merely average differences. Just as many women are taller than many men, many women are more competitive and more risk oriented than many men. Nonetheless, among those who are unusually tall or unusually competitive and risk oriented, males will be substantially overrepresented. Thus, while knowing an individual's sex tells us little about that individual's temperament, tastes, capacities, or achievements, average sex differences in these attributes will lead to average differences between the sexes as groups.

Biology and Public Policy

Some readers may already find themselves becoming uneasy. Many people are uncomfortable with biological explanations of social phenomena, believing such explanations to imply either the inevitability or the virtue of the status quo. For example, biologist Richard Lewontin and his colleagues have asserted that "the social function of much of today's science is to hinder the creation of [a just] society by acting to preserve the interests of the dominant class, gender, and race."[16] Law professor Herma Hill Kay has asserted that "sociobiologists use sexual difference as a natural evolutionary justification for continued female exploitation."[17] Others question the very legitimacy of the study of sex differences on the ground that such studies may reinforce existing political arrangements.[18] Implicit in these critiques is the mistaken view that there is some categorical hierarchy of explanations, under which those that incorporate biology are inherently inferior—either scientifically or morally—to those that ignore it.

To provide biological explanations is, some say, to confuse the *is* and the *ought,* apparently ignoring the fact that the primary function of science is to explain the natural world, not to justify or defend it.[19] If it is a fallacy to argue that the biological roots of a phenomenon demonstrate its desirability—as it surely is—then it is equally fallacious to infer that an argument for the biological roots of a phenomenon is implicitly an argument that the phenomenon is good. Put another way, it is more commonly the opponents than proponents of biological explanations who draw inferences of value from assertions of fact.

The discovery that all manner of social ills, from rape to child abuse, derive in part from biological predispositions does not justify these behaviors, especially since our desire to prevent those ills probably stems from the same kind of predispositions. Nor does the biological argument imply that attempts to reduce the prevalence of such behaviors are necessarily doomed, for a central finding of evolutionary psychology is the flexibility (albeit along predictable lines) of human behavioral responses. Nonetheless, an understanding of the deep origins of the phenomena of interest may assist us in an understanding of existing patterns as well as in developing designs for change.[20]

Some believe that we should not base law and policy on one or another specific view of human nature, but what is law but an attempt to respond to and shape human behavior? And what could be more essential to the shaping of human behavior than an understanding of the motivations behind that behavior? Law must always rest on an implicit view of human psychology. Legislators assume that passing a law will either increase or decrease the frequency of a particular behavior, an assumption that rests as much on a particular view of human psychology as the assumption that rewarding a pigeon with food for pecking at a bar will increase the pigeon's pecking. These assumptions necessarily rest on a theory of the mind. One who seeks to influence human behavior while remaining completely unconcerned with human motivations may end up selecting an appropriate policy option, but it will be only by accident.

Architects of current social policy relating to sexual equality often do not articulate it, but they also have a view of human nature and of the mind—the standard social-science view that humans, unlike all other mammals, possess a sexually monomorphic mind. Thus, they begin (and often end) with the assumption that men and women have the same preferences, abilities, and priorities, and that they will respond to economic and noneconomic incentives in the same way unless prevented by outside forces from doing so. In sex-discrimination lawsuits, for example, statistical methods of proof rest on the explicit assumption that in the absence of discrimination there would be equality in outcomes. The question for policy makers is thus not *whether* they should base policy on a particular view of human nature, but rather it is a question of *which* view of human nature they should embrace.

In deciding among competing explanatory theories, advocates of the "socialization" view have been very effective in imposing the burden of proof on those arguing for the importance of biological influences. Sometimes imposition of that burden is only implicit. Thus, in ordinary conversation an assertion that some sex difference is caused by discrimination or differential socialization typically passes without comment. If a biological cause is suggested, however, a demand for proof is likely to be immediately forthcoming and skepticism is expressed about such issues as whether relevant experiments were adequately controlled. Some critics are more explicit and expressly argue that those grounding their arguments in biology should bear a heavier burden than those ignoring biology.[21] Yet, it is unclear why this should be so. Both social and biological explanations are interpretations of causes of human social behavior and, one would think, subject to the *same* standards of proof. Indeed, a reasonable argument can be made that those denying biological contributors to sex differences should bear the burden of proof because of the inherent implausibility of their argument, since all other mammals exhibit such differences. Although a lower standard for biological arguments is not advocated here, an equivalently skeptical approach should be applied to sociological explanations.

Even if an understanding of the origins of sex differences is necessary to formulate coherent policies, we should not expect too much from science. Even a strong consensus on the nature of man and woman does not foreordain consensus on policy responses. Preexisting values are important and are not derivable from scientific fact. Thus, liberals, conservatives, radicals, and libertarians will probably favor liberal, conservative, radical, and libertarian policies despite their agreement on the underlying science. The burden of this book is to promote a more productive discussion of something that most people desire in the abstract but about which there is an extraordinary lack of definitional consensus: sexual equality.

I How the Sexes Differ

Workplace differences can be understood only against the backdrop of the important psychological differences between males and females. Males and females exhibit average differences in temperament, with males tending to be more aggressive and competitive on a host of measures and tending to engage in more dominance-seeking and risk-taking activity and females tending toward a more nurturant and social orientation. The sexes also exhibit average differences in cognitive ability, with males having special strength in mathematical and spatial domains, and females excelling in verbal domains. The combination of sex differences in temperament and cognition, as we will see later in part II, leads the sexes to pursue different occupations and to have somewhat different priorities.

The emphasis on differences between the sexes is not intended to suggest that women have none of the stereotypic qualities of men, or vice versa, only that they have them, on average, in lesser amounts. Thus, knowing that a person is a man may not tell you much about whether he is an aggressive risk taker, but knowing that one large group of people is made up of men and another is made up of women allows a confident prediction that the former group will have higher levels of risk preference and aggressiveness than the latter. Similarly, knowing that a person is a woman does not say much about whether she is strong in verbal abilities, but knowing that a broadly representative group is made up of women would suggest that the group will have verbal abilities superior to those of a broadly representative group made up of men.

If it seems that an inordinate focus of this and subsequent sections is on traits that men have "more" of or are "better" at, that is because of the nature of the questions that are being addressed, not because there is any implication that men are "better" than women, a position that would be purely nonsensical in any event. The question we are examining is why women do not have the same occupational outcomes that men do, which requires a look at what men do to achieve these outcomes and then to compare women along those dimensions. Although some may complain that men should not be the standard of comparison, such comparisons are demanded by complaints that women do not get what men get.

2 Sex Differences in Temperament

The entire life strategy of males is a higher-risk, higher-stakes adventure than that of females.

—Richard Alexander, *Darwinism and Human Affairs*

The sexes differ along a number of temperamental dimensions that will ultimately have workplace implications. Males from a very young age exhibit more aggressive, assertive, and competitive behaviors, most of which tend to be oriented toward achieving dominance over others (especially other males). They also tend to take more physical and nonphysical risks, while females tend to be more risk averse.

Females, also from a young age, tend to exhibit more nurturing behavior than males, paralleling the substantially greater maternal than paternal involvement in child-care activities worldwide. This pattern reflects the greater "person" orientation of females, in contrast to the greater "thing" orientation of males.

Men also differ from women in their greater willingness to pursue short-term mating strategies. Men have a greater desire than women for sexual variety, are willing to engage in sexual intercourse after a shorter period of acquaintance, and relax their standards for short-term mates considerably more than women do.

Suppose you were a visitor to a previously unknown society and you gave members of that society a test requiring them to identify the sex described by a series of adjectives. One sex is described as sentimental, submissive, superstitious, affectionate, dreamy, sensitive, attractive, dependent, emotional, fearful, softhearted, and weak; the other is described as adventurous, dominant, forceful, independent, strong, aggressive, autocratic, daring, enterprising, robust, stern, active, courageous, progressive, rude, severe, unemotional, and wise. What is the likelihood that the members of this society will identify the former sex as male and the latter as female? A cross-cultural study of sex stereotypes conducted by John Williams and Deborah Best suggests that the answer is "virtually zero." Respondents in at least twenty-three of the twenty-five countries studied associated each of the first set of adjectives with females and the second set with males.[1]

We have all grown up with generalizations, or "stereotypes," about males and females, but we are often told that there is something "bad" about stereotypes: they are said to be inaccurate, to constitute rigid all-or-none beliefs, and to cause people to ignore individuating information. In fact, however, existing data undermine all three of those assertions. Most studies of stereotype accuracy have shown them to be accurate generalizations; they tend to be probabilistic, rather than categorical beliefs such as "all Xs are Y"; and people tend to rely on stereotypes when they lack individuating information but attend to individuating information when it is present.[2] Indeed, stereotypes serve the essential function of allowing us to generalize from patterns of information.[3]

There is overwhelming empirical evidence of sex differences in temperament. In a review of sex-differences research over the preceding two decades, psychologist Alice Eagly found that existing research refutes four commonly asserted claims about sex differences: that they are small, inconsistent from study to study, artifactual, and inconsistent with stereotypes. Eagly noted that despite frequent repetition of such claims:

> It is not cultural stereotypes that have been shattered by contemporary psychological research but the scientific consensus forged in the feminist movement of the 1970s. Perhaps the idea that people's ideas about the sexes would be very misguided probably never should have seemed so plausible to psychologists, given the large amount of information that people process about women and men on a daily basis.[4]

Eagly's latter point is an important one. It is understandable that people might labor under false beliefs about things that they cannot directly observe. People in Bolivia may harbor strong and inaccurate views about Eskimos, for example, but, if so, these views were likely to have been acquired without their holders' ever having met one. On the other hand, members of the two sexes interact constantly. The notion that people consistently ignore a lifetime of personal experience to embrace unfounded myths about the sexes simply defies belief.

Aggressiveness, Dominance Assertion, Competitiveness, Achievement Motivation, and Status Seeking

The sexes are consistently found to differ in a constellation of related traits: aggressiveness, dominance assertion, competitiveness, achievement motivation, and status seeking. Each of these terms has its own somewhat different definition, but the traits are highly correlated and often overlapping.[5] Although "aggressiveness" may be defined narrowly to mean the infliction of harm on another,[6] a broader definition of the term encompasses most of the traits listed above. Thus, when we speak of an "aggressive soccer player" or "aggressive businessman," we are not necessarily describing a person who wishes to inflict harm on another, although we may be describing a willingness to step on competitors—literally in the former case and figuratively in the latter—in pursuit of his goals. Although evidence for sex differences in harm-inflicting behavior—the narrow definition of "aggressiveness"—is compelling, at least when focusing on direct verbal or physical aggressiveness,[7] it is not clear that such differences have much impact on the workplace outcomes of concern in this book and will therefore be largely ignored.

Competitiveness

Competition seems to come more easily to males than females and to be a more unalloyed positive experience for males.[8] Competition significantly increases the intrinsic motivation of men, while it does not do so for women.[9] A perception that an academic program is competitive may result in poorer performance by women but better performance by men.[10] Women also report higher levels of stress attendant to competition.[11]

Sex differences in competition appear in early childhood. Even preschool boys engage in more competitive activities than girls, activities that seem to elevate levels of adrenaline and noradrenaline.[12] A study of second through twelfth graders found that girls reported more-positive attitudes toward cooperation in school and less-positive attitudes toward competition in all grades than did boys.[13]

Sex differences in attitudes toward competition are reflected in children's play styles, as demonstrated in a well-known study of play by Janet Lever.[14] Boys were much more likely to engage in "games" (competitive interactions with explicit goals), while girls engaged in more "play" (behaviors with no explicit goal and no winners). Even when both sexes played games, the games were of different types. Games like hopscotch and jump-rope are "turn-taking" games, in which any competition that exists is indirect. When boys competed, they were more likely to compete head-to-head in zero-sum competition. Because boys cared more about being declared the winner, their games were usually structured so that there would be a clear and definite outcome.

The boys in Lever's study seemed to display a more instrumental approach to competition. Exhibiting a difference that may have later workplace implications, boys were much better than girls at competing against friends and cooperating with teammates whom they did not like. Anecdotal accounts of childhood play suggest that when boys pick teams for games they pick the people they believe to be the best players irrespective of whether they like them, while girls are more likely to choose their friends irrespective of skill. Although Lever observed that among boys there were repeated disputes over the rules, no games were terminated as a consequence of these disputes. In fact, she concluded, boys enjoyed the rule disputes as much as the game. In contrast, quarrels among girls over application of rules were likely to terminate the game, perhaps anticipating what primatologist Sarah Hrdy has referred to as "the well-documented problem that unrelated women have working together over a long period of time."[15]

Competition is simply a greater part of male life, even among children. If one boy tells another boy that he can spit ten feet, the response is likely to be either "I bet you can't" or "I can spit farther"; if a girl tells another girl that she can spit ten feet, the response is likely to be either "So what?" or "That's gross." Psychologist Eleanor Maccoby observes that "even when with a good friend, boys take pleasure in competing to see who can do a task best or quickest, who can lift the heaviest weight, who can run faster or farther."[16] A study of free-play activities in fourth and sixth graders found that boys were engaged in direct competition with other boys 50 percent of the time, while among girls direct competition occurred only 1 percent of the time.[17]

This is not to say that girls do not compete, but they often employ different means and pursue different ends. Sarah Hrdy attributes the belief that females are less competitive than males to the failure of scientists to examine "women competing with one another in the spheres that really matter to them."[18] For purposes of this book, that is a critical point. One need not reach global conclusions about whether men are more competitive than women in all spheres; instead, the question is whether men are more competitive in the extra-domestic hierarchical settings of the workplace.

Dominance Assertion

Dominance assertion is, in a sense, a form of competition. Dominance behaviors are those intended to achieve or maintain a position of high relative status—to obtain power, influence, prerogatives, or resources.[19] When children get together, even in the preschool years, dominance hierarchies emerge spontaneously: some children are more influential and less subject to aggression by others. Boys engage in a significantly greater amount of dominance-related play than girls, such as playing with weapons and engaging in rough-and-tumble play.[20] In mixed-sex groups in nursery schools, boys end up disproportionately at the top of the hierarchy within the classroom.[21] Even among preschoolers, the expression by girls of their ideas seems to be significantly curtailed in the presence of boys.[22]

Although boys generally assume the dominant positions in mixed-sex hierarchies, boys and girls mostly establish separate hierarchies, a division that occurs in large part because boys and girls spend most of their time with others of the same sex. One of the most robust sex effects of early childhood is the powerful tendency that young children exhibit toward sex segregation. Up until about age two, children show little preference regarding the sex of playmates, but during the third year, same-sex preferences begin to emerge in girls, followed perhaps a year later by a similar preference among boys.[23] Although girls show an earlier same-sex preference, when the preference emerges in boys it is stronger.[24] After about age five, boys take a much more active role in policing the boundaries of sex-appropriate behavior. These findings have been replicated in a wide range of cultures.[25]

After the preference for same-sex playmates develops, children continue to play with members of the opposite sex, especially if required to by adults or by the dearth of same-sex playmates. When children find themselves surrounded by large numbers of children of their own age, however, the impulse to self-segregation is strong. As children move into later stages of childhood, an increasingly large portion of their time is spent with same-sex others, despite pressures by teachers to require greater sexual integration. Eleanor Maccoby has suggested that part of the aversion that girls have to boys is that girls find it difficult to influence the boys.[26] Jacklin and Maccoby found, for example, that among unfamiliar pairs of thirty-three-month-old children, boys were less likely to pay attention to instructions from girls than girls were to those from boys.[27]

In their same-sex groups, both boys and girls establish hierarchies, but the hierarchies differ in strength and in the traits that lead to dominance. Boys' dominance hierarchies tend to be more stable and well defined than those of girls.[28] That is, the boys largely agree about who is on top, and these rankings tend to persist over time. Among girls, on the other hand, hierarchies are more fluid and there is considerably less agreement concerning the relative rankings of individual girls. Hierarchies among boys tend to be established quickly, often during their first interaction.[29] Among boys, the critical determinant is "toughness," both physical strength and unwillingness to back down. Among girls, however, status tends to be achieved through physical attractiveness and friendship with popular girls.[30]

Important sex differences in dominance behaviors and aggressiveness are obscured by looking only at frequencies, since the sexes also differ in the types and causes of these behaviors. Psychologists Martin Daly and Margo Wilson, for example, have found a consistent worldwide pattern: homicides tend be committed by and against unmarried young males. Many are what Daly and Wilson label "trivial altercations," either "escalated showing-off disputes" or "disputes arising from retaliation for previous verbal or physical abuse." Although to an outside observer the precipitating event may seem trivial, many of these disputes are "affairs of honor," in which the precipitating event "often takes the form of disparagement of the challenged party's 'manhood': his nerve, strength or savvy, or the virtue of his wife, girlfriend, or female relatives."[31] Failure to respond to challenges to reputation or signs of disrespect leads to loss of face and of relative status.

The foregoing does not mean that women are not interested in achieving status. However, the route to female status, across history, has been quite different for men and women. Men have generally achieved status through dominance over other men; women have achieved status not primarily by achieving dominance over other women (or over men) but rather through their association with high-status men.

Achievement Motivation and Response to Failure

Although in many respects the sexes are similar in their motivation to achieve, some sex differences are consistently reported, many of which are associated with attitudes toward failure. For example, when given a choice of tasks to perform, males are more likely to select the more difficult task and females the easier one.[32] Females tend also to be more adversely affected by failure and more likely to give up than males,[33] and they are somewhat more likely to attribute failure to lack of ability rather than lack of effort.[34] Males, on the other hand, are more likely than females to improve in performance after failure.[35]

Confidence is an important contributor to achievement motivation, and competition seems to exaggerate sex differences in confidence. Anticipation of competition results in lower confidence levels in females than males.[36] Moreover, females' performance predictions tend to be relatively unstable and subject to change with single encounters, while males are less likely to allow one failure to diminish their performance expectations.[37]

Females seem to have a greater need than males for feedback about performance in order to achieve or maintain high levels of self-confidence in their performance capabilities.[38] Such feedback seems to be more central to women's self-esteem than is the case for men, who exhibit a weaker relationship between positive feedback and global self-esteem.[39] As we will see later, this need may contribute to the dissatisfaction that women who study science often feel in moving from high school to college, when they leave an environment in which they are lavished with attention and praise and enter an environment in which they join the relatively anonymous masses.

Competition and Dominance in Mixed-Sex Groups

Studies measuring only attributes of individuals often miss important dynamics of social interactions. Males and females sometimes exhibit different competitive and dominance behaviors in same-sex and mixed-sex groups. Within same-sex groups, for example, males engage in more dominance behaviors than females do. This tendency is moderated to some extent in mixed-sex groups, with male dominance behaviors tending to abate somewhat and female dominance behaviors to increase; still, males in mixed-sex groups continue to act more dominantly.[40] Not surprisingly, in same-sex pairings, a high-dominance individual will assume a leadership role over a low-dominance individual. However, when a high-dominance woman is paired with a low-dominance man, the low-dominance man tends to assume the leadership role. It is not that he asserted dominance over the woman to become the leader, but rather that the dominant woman selected him to be the leader.[41] Perhaps this represents the same phenomenon as the many examples of very able women pushing their somewhat less able husbands to be more successful than the husbands would have been on their own.

Mixed-sex competition seems to be a quite different experience for its participants from same-sex competition. A study comparing responses of competitors to male, female, and machine opponents in a video game found some interesting patterns. First, it appears that men do not like to compete against women. Men who were measured as being low in competitiveness demonstrated low physiological arousal when their competitor was a woman, apparently because they were less engaged in the competition than when their opponent was either a machine or a man. This result is consistent with suggestions that some men may reduce effort in situations in which they are concerned about being outperformed by women.[42] High-competitive men, on the other hand, demonstrated the highest level of engagement and the highest level of negative affect at the conclusion of the competition when their opponent was a woman. Why should men feel discomfort when competing against a woman? One potential reason is that men may feel they will be criticized for playing to win against a woman.[43] Another reason is that men may perceive these encounters as "no-win" situations for them, because they do not get the same credit for winning if their opponent is "just a woman," but losing to a woman is worse than losing to a man.

The literature dealing with female performance in mixed-sex competition is not entirely consistent, but psychologist Carol Weisfeld summarizes the literature as revealing that "it is almost always the case that some females will depress their performance levels, resulting in victory for males."[44] This female suppression of effort does not appear to be consciously motivated. Females who perform at their highest level against males tend to be those characterized by masculine or androgynous temperaments. For them, competition against males appears a positive experience, perhaps for precisely the opposite reason that competitive males do not like the same competition: if she wins, the credit for beating a male is high, but if she loses, there is no shame in losing to a male.

Head-to-head competition between the sexes for status in hierarchies is a relatively recent phenomenon. Humans, like other primates, may not be "wired" in such a way as to evoke competitive responses from members of the opposite sex. Thus, perhaps it should not be surprising that neither males nor females view competition with the opposite sex as equivalent to same-sex competition.

Risk Taking

As with the other traits considered in this chapter, the sexes differ in risk taking from childhood.[45] One of the best measures of physical risk taking in children is the incidence of accidental death and injury. In most industrialized countries, including the United States, accidents are the leading cause of death for children older than one year. A World Health Organization study of accidental-death rates in fifty countries found a substantially higher rate for boys in all countries, with a ratio of male to female deaths of 1.9:1 in Europe and 1.7:1 in non-European countries.[46] Notwithstanding greater equality and socially sanctioned androgyny, the male/female accidental-death ratio actually increased in the United States from 1960 to 1979.[47]

Greater risk taking among boys is a robust finding.[48] Boys are exposed to greater risks not only because they are more likely to engage in risky behaviors, but also because when engaging in the same activity as girls, they are more likely to perform it in a risky manner. Boys are substantially more likely to approach hazardous items than girls, and they differ in how they approach them, with girls tending to look and point and boys tending to touch and retrieve them.[49]

Several factors appear to account for the greater inclination of boys to engage in risky activity. Boys tend to have both a higher activity level and poorer impulse control, both traits that are associated with injury rates.[50] Three factors are correlated with self-reported risk taking in both boys and girls: attribution of injuries to bad luck, a belief that one is less vulnerable to injury than one's peers, and downplaying the degree of risk. Boys score higher than girls on all three of these traits.

Boys are less likely than girls to abstain from a risky activity simply because they have seen a peer injured while engaging in the same activity. The best predictor of girls' willingness to take a particular risk is their belief about the likelihood of getting hurt, while for boys it is the perceived severity of the injury.[51] That is, girls tend to avoid risks if they think they might get hurt, while boys seem to be willing to take risks if they do not think they will get *too* hurt. It is possible, however, that positive attitudes about risk may result in part from risk taking, rather than causing it.[52]

In adolescence and adulthood, sex differences in risk taking increase. Men are disproportionately involved in risky recreational activities such as car racing, sky diving, and hang gliding. Indeed, sex is the variable most predictive of the extent of participation in high-risk recreation.[53] The driving style of men also shows a greater propensity toward risk.[54] Men are disproportionately represented in risky employment, as well. Over 90 percent of all workplace deaths in the United States are males.[55] A list of dangerous occupations is a list of disproportionately male occupations: fisherman, logger, airplane pilot, structural metal worker, coal miner, oil and gas extraction occupations, water transportation occupations, construction laborer, taxicab driver, roofer, and truck driver.[56]

Men's greater propensity to risk their lives is demonstrated by a study of the recipients of awards granted by the Carnegie Hero Fund Commission. Of the 676 acts of heroism recognized from 1989 through 1995, 92 percent were performed by males.[57] Moreover, over one-half of those rescued by women were known to the rescuer, while over two-thirds of those rescued by men were strangers. Although this is not a random sample of heroes, since one must be nominated for the award, it is likely that, if anything, the sex difference is understated because acts of heroism by women would tend to attract more attention than those by men.

Risk taking is statistically correlated with a number of other stereotypically male traits. People who rate high on achievement and dominance, for example, tend to be high risk takers.[58] Risk taking and competitiveness may be related, since competition-prone individuals tend to be willing to take greater risks in pursuit of their competitive objective.[59] High risk takers also fight more frequently, are more socially aggressive, take more dares, and participate in more rough sports and physical activities such as hunting, mountain climbing, and auto racing. In contrast, risk taking is negatively associated with a number of stereotypically feminine traits: affiliation, nurturance, succorance, deference, and abasement.

Psychologist Elizabeth Arch has suggested that the sex differences in achievement-orientation previously discussed may be explained in part by sex differences in risk taking. From an early age, females are more averse not just to physical risk but also to social risk, and they "tend to behave in a manner that ensures continued social inclusion."[60] This aversion to risk may be partially responsible for women's disproportionately low representation in positions involving "career risk," which may adversely affect their prospects for advancement. This pattern suggests that what is sometimes labeled women's "fear of success"[61] is in fact the more prosaic and easier-to-understand "fear of failure."

One's willingness to take risks depends in large part upon the relative values that one places on success and failure. A person whose appetite for success exceeds his aversion to failure will be inclined toward action; a person whose aversion to failure exceeds his appetite for success will be inclined not to act.[62] A strong motive to achieve or to avoid failure may also bias the actor's subjective probability of outcome. That is, an achievement-oriented person may have a higher expectation of success than is objectively warranted, while a person with a high motivation to avoid failure may consistently underestimate the chance of success.[63]

Nurturance, Empathy, and Interest in Children

Females in all known societies exhibit more nurturing behavior than males both inside and outside the family. Everywhere it is women who are the primary caretakers of the young, the sick, and the old.[64] Even among young children, girls tend to exhibit more nurturing behavior, and throughout the adolescent years, girls have a greater preference than boys for more caring, personal values.[65] Girls' interest in infants increases substantially at menarche.[66]

Just as the description of greater male aggressiveness, dominance assertion, and risk taking does not mean that women lack these traits altogether, to say that women seem to be higher in nurturance and empathy does not mean that men lack these traits or that they are not capable of caretaking behavior. Men are certainly as capable as women of learning many of the routine behaviors of parenthood, such as changing diapers and making dinner.[67] The important question, however, is whether the connection to the infant and attunement to its state is the same for men as for women. There is some evidence to suggest that women's nurturant responses have a stronger physiological underpinning than men's.[68]

Sex differences in parental care are universal across cultures. While the level of paternal involvement varies considerably among societies, there is no society in which the level of direct paternal care approaches that of mothers. Among the Aka pygmies of Central Africa, fathers provide more direct care to their children than in any other society known to anthropologists. Nonetheless, Barry Hewlett found that males, on average, hold their infants for a total of 57 minutes per day, compared to 490 minutes for mothers.[69] The greater female contribution to child care in contemporary Western society is, of course, well known. The disparity between male and female care is exaggerated after divorce, with noncustodial mothers being far more likely to maintain close contact with their children than noncustodial fathers.[70]

Psychological studies generally confirm that women are more empathic than men, in the sense that they experience a "vicarious affective response to another's feelings." As psychologist Martin Hoffman observed in 1977, the most striking feature of the empathy findings in a whole host of studies "is the fact that in every case, regardless of the age of the subjects or the measures used, the females obtained higher scores than did the males."[71] While some subsequent studies have found no sex difference, they tend to be studies measuring the ability to identify other people's feelings rather than measuring the subject's emotional reaction to those feelings.[72] Personality studies also show a substantial correlation between nurturance and empathy.[73]

The more social orientation of females is reflected in a consistently found sex difference in "object versus person" orientation. Females of all ages tend to be "person oriented," while males tend to be more "object oriented."[74] As early as the first year of life, girls pay more attention to people and boys pay more attention to inanimate objects. In a study of college students, male and female subjects were shown a series of pictures of human figures and mechanical objects in a stereoscope so that a picture of a human figure and a picture of a mechanical object were falling on the same part of the subject's visual field. The theory behind the experimental design is that when two stimuli are competing, subjects will pay attention to the stimulus that is more interesting to them. Males had a greater tendency to report seeing the objects, while females tended to report seeing the human figure.[75]

Differences in orientation affect the way people perceive themselves. Women's self-identity and self-esteem tend to be centered around sensitivity to and relations with others, while men's self-concepts tend to be centered around task performance, skills, independence, and being "better" than others.[76] In one study, 50 percent of the women but only 15 percent of the men agreed with the statement, "I'm happiest when I can succeed at something that will also make other people happy."[77] As psychologist Carol Gilligan has argued, "Women not only define themselves in a context of human relationship but also judge themselves in terms of their ability to care."[78]

Sex Differences in Sexuality

The stereotype of men as being more indiscriminate in their sexual behavior also has a solid foundation, and, as we will see in chapter 13, this difference has implications for sexual harassment in the workplace. A substantially larger portion of men's mating activity is oriented toward short-term mates. Men have a greater desire for sexual variety than women do, and they are willing to have sexual relations with a new partner after a shorter period of acquaintance. Men also relax their standards for short-term mates considerably more than women do.[79]

By virtually any measure, men's sexuality is "closer to the surface" than women's. This is not to suggest that women do not enjoy sex—or even that they do not enjoy it as much as men—but as a rule, men and women differ in the amount that they think about sex and devote effort to obtaining it, as well as in their threshold for engaging in sexual behavior.[80] A large-scale survey of American sex practices found that although men and women do not differ in how appealing they find the idea of sexual intercourse, 54.4 percent of men say that they think about sex at least once per day compared to only 19.7 percent of women, and 26.7 percent of men, compared with only 7.8 percent of women, reported that they masturbate at least once per week.[81]

A graphic demonstration of greater male willingness to participate in casual sexual encounters was provided in an experiment conducted by Russell Clark and Elaine Hatfield. Male and female volunteers approached members of the opposite sex and asked them one of three questions:

(1) whether they would like to go out on a date;
(2) whether they would like to go to the volunteer's apartment; and
(3) whether they would like to have sex.

Although male and female subjects were equally likely to respond favorably to the request for a date, men were far more likely to respond positively to a request to visit an apartment or to have sex. In fact, approximately three-quarters of the men responded positively to the invitation to sex (a substantially higher proportion than were willing to go on a date), while not a single woman responded positively.[82] Thus, it appears that many men are willing to have sex with women they would not even be willing to date.

Some have argued that the explanation for this sex difference may be women's fear for personal safety rather than a lesser inclination to engage in casual sex.[83] While the possibility that some women might have been receptive to a solicitation for sex if adequate guarantees of safety had been provided cannot be completely discounted, it seems unlikely that the fraction would have approached three-fourths. A large-scale survey of sexual practices in America found that 33.8 percent of male respondents aged 18 to 44 found the idea of sex with a stranger very or somewhat appealing, while only 10.4 percent of women did so,[84] a much less imbalanced ratio than found by Clark and Hatfield, but a substantial difference nonetheless. (It is unfortunate that the sex survey used such broad age groupings, as one would expect eighteen-year-olds and forty-four-year-olds to have very different views on sexual matters.)

Is it true that many men are actually willing to have sex with women they would not be willing to date? Some clue to that question is provided by the number of men who visit prostitutes. According to the same survey on sexual practices, approximately ten times as many men as women (16.8 percent versus 1.6 percent) responded positively to the question whether they had ever had sex with a person they paid or who had paid them for sex.[85] Although the survey does not reveal the breakdown of those who paid versus those who were paid, it seems a reasonable inference that the overwhelming number of men who responded positively had paid, while most of the women were paid. It also seems a reasonable inference that most of the men would not be willing to date the prostitutes they visited.

Comparisons of the number of sex partners reported by men and women usually find that men report more partners than women. Because the total number of sex partners must be the same for men and women (in a heterosexual population), it has been argued that the disparity in reported partners is a consequence of male overreporting and female underreporting, both due to respondents' providing socially desirable responses. A recent study by Devon Brewer and his colleagues found support for a different explanation—that there is a relatively small group of women who have a large number of sexual partners but who are undersampled in the typical survey. By adjusting for prostitution-related sex, they found that the apparent sex disparity in the number of sex partners disappeared.[86]

The largest differences between men and women in mating behavior relate to the short-term matings described above, where men have substantially lower standards than women.[87] Men are willing to have sexual relations with women after a much shorter period of acquaintance, and are generally much less picky about a wide variety of mate attributes—age, personality, and sexual promiscuity. Indeed, promiscuity can be a somewhat positive trait in a potential short-term partner, because it signals availability, though it is perceived very negatively in a potential long-term mate. In contrast, women's partner preferences differ less between short-term and long-term potential mates.

This chapter has revealed that males and females differ, on average, in a number of temperamental traits. Men tend toward competition, women toward cooperation. Men seek to achieve dominance over others, while women seek to cement social relations. Men tend to be object oriented, while women tend to be person oriented. Men tend to be eager to engage in short-term sexual relationships, while women tend to be more reticent. These are just statistical generalizations, but they hold true not just in our society but also cross-culturally. As we will see in the next chapter, these temperamental sex differences are accompanied by sex differences in cognitive traits as well.

3 Sex Differences in Cognitive Abilities

The sexes differ in performance on many cognitive tasks, although they differ little, if any, in general cognitive ability. Many tests of spatial ability, especially mental rotation, show a consistent male advantage, while others, such as object location, show a female advantage. These patterns are reflected in sex differences in navigation, with men being more likely to rely on cardinal directions and women being more likely to rely on landmarks.

Males have only a modest advantage on tests of mathematical ability in broadly representative samples. At the highest levels of mathematical performance, however, a substantial sex disparity appears as a consequence of the somewhat higher male mean and greater male variability. The male advantage is exhibited primarily on tasks that measure concepts and reasoning. In contrast, females outperform males on tests of computation. A large sex difference is also found on mechanical reasoning, although the effects of experience can be difficult to disentangle.

Females have an advantage in some verbal abilities—principally spelling, grammar, and "verbal fluency"—in broadly representative samples. This advantage exceeds in size the male math advantage. In select samples, however, there is less difference between males and females in verbal ability.

Evidence for a decline in sex differences in cognitive abilities over time is equivocal. At least on tasks for which the most reliable sex differences are found—mental rotation, advanced mathematics, spelling, and verbal fluency—the differences have been relatively stable. Changes in tests, in populations sampled, and in publishing conventions, however, make it difficult to reach firm conclusions.

The sexes differ not only in their temperaments but also in their cognitive abilities.[1] Certain commonly accepted generalizations about the sexes are not correct, however, or at least must be substantially qualified. Although it is often said that men are better at spatial and mathematical tasks and women are better at verbal tasks, the truth is more complex. Men are better at some spatial and mathematical tasks, while women are better at others, and no sex differences are found in others.[2] Similarly, women are better at some verbal tasks, while men are better at others, and for yet others there are no sex differences at all.[3]

Measurement and Reporting of Differences

Before we turn to specific sex differences, a brief explanation of the measurement of such differences is in order. When researchers talk about "significant" sex differences, they might mean one of two things. They might mean that the difference between male and female performance is "statistically significant," meaning that the observed difference between male and female samples—no matter how small—can be attributed to reliable differences between the sexes (or the subgroup of the sexes from which the sample is drawn, such as college applicants) rather than being simply a product of chance. When sample size is very large, even tiny differences between two groups may be statistically significant, because large samples tend to differ little from the population from which they are drawn. Therefore, with large samples, differences that are too small in magnitude to have any meaningful real-world effect may nonetheless be labeled "significant."[4]

A concept that can provide greater insight into the magnitude, and therefore the practical meaning, of differences is *effect size*. Effect size (denoted as *d*) is a measure of the difference between the means of groups, expressed in terms of the number of standard deviations separating the two means.[5] In sex-differences research, *d* is usually calculated as the male mean minus the female mean divided by the pooled standard deviation. An effect size of 1.0 indicates that the male mean exceeds the female mean by a full standard deviation, which in practical terms means that the average male exceeds the performance of 84 percent of females, assuming that the two groups are equally variable. Effect size, it should be noted, is not an alternative to statistical significance. In order for a given effect size to be viewed as "real" and not an artifact of chance, the difference between the means of the two groups must also be statistically significant.

The proportions described above would be different if one group is more variable than the other.[6] On most cognitive measures—especially ones that favor males—male performance is more variable than female performance.[7] If the male and female means are identical but males are more variable than females, then at both the high and low ends of the distribution, males will outnumber females.

Different characteristics of the male and female distributions are relevant to different questions. For example, if we want to predict whether a male or female chosen at random would be better along some dimension, we would care primarily about group means. If the effect size is zero, there would be no reason to think that a male chosen at random would perform better—or worse—than a female chosen at random, regardless of any sex difference in variability.

Sometimes, however, we are interested primarily in those who are at the extreme end (usually the high end) of ability. For example, if we want to investigate the extent to which sex differences in mathematical ability are responsible for sex differences in representation in math-intensive occupations, we care not about the center of the curve but rather the high end. Even on tests for which the female mean exceeds the male mean, there may be no sex difference among the very able. For example, female twelfth graders outperform males on spelling ability on the Differential Aptitude Test (DAT), with an effect size of -0.50. This means that only about 30 percent of males score above the female mean. However, because of greater male variability, males and females are equally represented in the top 1 percent of performance.[8]

Because of sex differences in variability on many tests, broad inferences about males and females in general cannot properly be drawn from samples that are not representative of the entire population. Assume, for example, a trait for which the overall population d is zero, but for which there is greater male variability. A sample taken from the middle of the ability distribution will show no average sex difference. However, a sample taken from the high end of ability—for example, from the college-bound—will show a male advantage, because of the greater number of males in the high end of the distribution. Similarly, a sample taken from the low end of the sample will show a female advantage, because of the greater number of males in the low end of the distribution.[9]

Sex Differences in Specific Cognitive Traits

The fact that sex differences in specific cognitive traits are reliably found does not suggest that one sex is, overall, more intelligent than the other. On average, males and females differ with respect to certain traits, but these differences seem to average out. While it is impossible to reach any firm conclusions about sex differences in intelligence from standard IQ tests, since they are deliberately constructed to minimize sex differences, tests that have not been so constructed but that load heavily on g (the general intelligence factor) reveal no average sex differences.[10] On the other hand, the greater variability of males means that in both the upper and lower portions of the curve, males predominate. For example, there are approximately 20 percent more males than females among those with an IQ over 140.[11]

Spatial Ability

Tests of spatial ability tend to show the most reliable sex differences favoring males. Among the spatial tasks favoring males are mental rotation, spatial perception (field independence), spatial visualization, and targeting.[12] The largest and most consistent of these differences is on mental-rotation tests, which require the subject to imagine what a figure would look like if rotated in a particular way. A meta-analysis of 286 effect sizes from a large number of studies found an average effect size of 0.66 for adults, though the effect size in many studies approaches or exceeds 1.00.[13]

As an illustration that "spatial ability" is not a unitary concept, females do better than males in "object location," again often yielding effect sizes of approximately 1.0.[14] Psychologists Irwin Silverman and Marion Eals have conducted a series of studies in which they show subjects an array of pictures of objects and then show them a second array in which some of the objects have changed location. Females are much better than males at identifying which objects have been moved.[15] This sex difference is late in emerging, with male performance being relatively constant from childhood to adulthood and female performance improving over that time.[16]

Sex differences in spatial ability carry over into the more ecologically relevant task of navigation. Men tend to be more attuned to compass directions, while women are more attentive to landmarks.[17] Researchers have found a huge difference between the sexes in navigation of a virtual maze, with an effect size of 1.59 for speed of completion and −1.40 for the number of spatial errors made.[18] Results on the maze test were highly correlated with mental-rotation scores, but not with tests of verbal ability. Similarly, Irwin Silverman and his colleagues at York University found substantial sex differences in actual outdoor navigation through a woods. They also found a significant correlation between direction finding and performance on a test of 3-D mental rotation, but not between direction finding and non-rotational spatial abilities or general intelligence.[19] A study recently conducted at West Point found similar sex differences in way-finding ability among cadets.[20]

The well-established sex difference in geographic knowledge[21] may be partially due to differences in spatial ability. James Dabbs and his colleagues found that geographic knowledge was independently correlated with both sex and spatial skills, leading them to conclude that "sex differences in navigation strategy and geographic knowledge are mediated in part, but not entirely, by sex differences in cognitive spatial skills."[22] Although the spatial link is interesting, the male superiority in geographic knowledge parallels the male superiority in most forms of "general information," such as history or current events, areas where a spatial contribution would not be expected.[23] Indeed, on all seventeen of the subject tests of the Graduate Record Exam (GRE), males outperform females, with effect sizes ranging from 0.16 for psychology to 0.76 for political science. The ratio of males to females among those scoring over 700 on the European History College Board test has ranged, over the years, from 4 to 1 to 6 to 1.[24]

Mathematical Ability

Sex differences in performance on standardized tests of mathematical ability also regularly appear. Males consistently excel in tests of mathematical reasoning, while females excel, although by substantially smaller margins, in arithmetic calculation.[25] In nationally representative samples, effect sizes for overall mathematical ability are not terribly large, mostly between 0.10 and 0.25.[26]

However, because of greater male variability, males outnumber females by almost 2 to 1 in the top 10 percent of math ability. As is typically the case, the difference is larger in more select samples. Thus, on the mathematics portion of the Scholastic Assessment Test (SAT-M), taken primarily by college-bound high-school students, the effect size in 2000 was 0.31.[27]

Sex differences at various levels of performance on the mathematics portion of the SAT (SAT-M) illustrate the combined effect of a higher male mean and greater male variability. Of those scoring 700 or over, the sex ratio is almost 2 to 1. The disparity can increase dramatically with increasing ability, however. In a sample of mathematically precocious youth (primarily seventh graders), Camilla Benbow and David Lubinski found that the male/female ratio of those scoring over 500 was 2:1; it increased to 4:1 at 600, and to 13:1 at 700.[28] Among all test takers, not just precocious ones, the ratio of males to females among those earning perfect scores is now just around 3:1. There is some reason to believe that the Educational Testing Service (ETS) is purposely shrinking this ratio, however, given the steady and rapid decline in the ratio since 1996: 1996 (3.38:1), 1997 (3.25:1), 1998 (3.05:1), 1999 (2.99:1), 2000 (2.84:1).[29] Given the large number of test takers (over one million per year), these are highly significant changes. Moreover, in light of the fact that course-taking differentials declined substantially earlier than this decline in test scores, not to mention the fact that the SAT-M is designed to be relatively curriculum-independent (it is designed to require only knowledge gained through grade nine),[30] there is no obvious answer other than that the ETS has been aggressive in weeding out questions on which boys do better than girls.

The ACT Assessment, the second most popular college entrance exam, tends to focus on curriculum-based knowledge rather than the verbal and quantitative reasoning tapped by the SAT. Even here, however, males predominate among the perfect scorers. The ACT has four parts: English, mathematics, social studies, and natural sciences. While girls outnumber boys by about 1.8 to 1 on the English test, boys outscore girls by substantially larger margins on the other tests (3.3:1, 2.3:1, and 5:1, respectively).[31]

Although males outperform females on tests of mathematical concepts, females outperform males on tests of computation. Thus, the greater a test's emphasis on mathematical concepts and reasoning, as opposed to arithmetic calculation, the greater the sex difference. These differences are confirmed by examination of the kinds of mistakes that males and females make in solution of math problems. Boys tend to make more errors in operations, such as adding, subtracting, borrowing, and carrying, while girls are more likely to select an inappropriate operation, such as dividing when they should multiply.[32] Further, girls are more likely than boys to be distracted by irrelevant information in solving algebraic word problems.[33]

In contrast to the consistent pattern of superior male performance on standardized math tests is the equally consistent pattern of superior female performance in coursework, as measured by grades. Girls from an early age get better grades not only in verbally challenging coursework, where standardized test results would suggest that they would excel, but also in math and science. A number of explanations for the disparity in performance between standardized tests and grades have been suggested.

Girls' greater writing and spelling skills are probably partially responsible for the disparity between grades and standardized-test performance, since superior verbal presentation may lead to better class grades among people who have the same command of the substantive material. Moreover, especially in primary and secondary school, grades are often based in large part upon perceived effort, completion of homework assignments, and compliance with teachers' demands, behaviors more characteristic of girls than boys.[34] Even among gifted children, Camilla Benbow reports that she encounters "many students who get A's on all tests but poor overall grades because they fail to turn in homework."[35] Mastery of course content is an important component of grades, but it is not the only one.

The tendency of girls to get relatively better grades than test scores has led some to suggest that tests are unfair to girls, but this is true only in a very limited sense. Girls' SAT scores slightly underpredict their future academic performance, but much of that effect disappears when difficulty of college coursework is controlled.[36] Reliance on grades alone in admissions decisions, however, would result in underprediction of *boys'* future academic performance.[37] A combination of tests and grades thus appears to yield the most accurate predictions for both sexes.

Mechanical Ability

Another dimension of sex differences is mechanical ability. On the Differential Aptitude Test, male twelfth graders substantially outperform females on mechanical comprehension, with an effect size of around 0.90.[38] Similar results ($d = 0.95$) have been obtained on the Air Force Officer Qualification Test (AFOQT), which is used in the selection of candidates to be U.S. Air Force officers.[39] In the top 10 percent of mechanical reasoning ability, males outnumber females by approximately 8 to 1.[40]

The extent to which this sex difference is due to innate differences in ability versus experience is difficult to disentangle. Unlike mathematical and verbal experience, which school curricula ensure are fairly equally distributed between the sexes, little of the school curriculum is oriented toward mechanical skills. As a result, mechanical experience is gained largely outside of school, and boys tend to engage in more mechanical activities than girls do. Among fifteen- to twenty-two-year-olds, the sex ratio in "auto and shop information" in 1980 was approximately 66 to 1 in the top 10 percent and 464 to 1 in the top 5 percent. Mechanical comprehension can be substantially increased with practice, but, in line with the notion that people's inherent capacities influence their environment, those who obtain the most practice are likely to be those with the greatest native ability.

Verbal Abilities

As with spatial abilities, the sexes have different strengths in the category of verbal abilities, with the overall female advantage beginning early in life.[41] Throughout the life course, females are better at spelling, grammar, and "verbal fluency,"[42] the latter term indicating the ability to produce words or sentences with particular constraints on them, such as listing words starting with the "ee" sound. Females also perform better on tests of verbal memory.

In broadly representative samples, females outperform males on a variety of verbal tasks. Among high-school students, for example, girls outperform boys on various measures of writing ($d = -0.57$), language use ($d = -0.43$), and reading ($d = -0.23$).[43] These effect sizes are greater than those favoring males for mathematical ability in national samples. Indeed, on the National Assessment of Educational Progress (NAEP), male eleventh graders score at about the same level as female eighth graders.[44] Females do not outperform males on all tests of verbal ability, however. In fact, the verbal portion of the Wechsler Adult Intelligence Scale (WAIS) shows small, but consistent, differences in favor of men, and women do not have larger vocabularies than men.[45]

In more select samples, the female advantage sometimes declines or disappears. For example, males consistently outscore females on the SAT-V, although by only a small amount (4 points in 1996), a fact that may be due to the greater number of girls who take the SAT-V or the test's heavy weighting of verbal analogies, a task on which males excel.[46] On the other hand, as we have seen, female perfect scorers on the English portion of the ACT outnumber males by 1.8:1.

Focus on a single dimension of cognitive ability may result in an underestimation of the practical effect of sex differences. A study measuring the effect of ethnicity, socioeconomic status, and sex on a broad range of mental abilities—spatial, verbal, mathematics, and memory—found that sex accounted for 69 percent of the variance in mental abilities, compared with 9 percent and 1 percent for ethnicity and socioeconomic status, respectively.[47] Referring to this study, psychologist Diane Halpern noted that it "seems that sex as a variable can explain a much greater percentage of the total variance when several sex-related cognitive abilities are considered simultaneously."[48]

Is the Sex Difference Declining?

It is often suggested that sex differences in cognitive performance are disappearing, and there has indeed been a decrease in the sex difference favoring males on standardized mathematics tests in nationally representative samples, though not in elite samples. There has been little or no decrease, however, in the overall sex difference favoring females in verbal abilities.

Several meta-analyses have examined the question of the possible decline in sex differences with varying results, but most studies do not support the view that differences are shrinking. The meta-analyses that have found the sex difference in spatial ability to be declining generally combine studies employing different spatial tests.[49] In a study of results on a single test, Alan Feingold found a decline from 1947 to 1980 in the effect size on the Space Relations subtest of the Differential Aptitude Test from 0.37 to 0.15, although, as Diane Halpern has pointed out, this is a test that does not reliably show sex differences anyway.[50] Mary Masters and Barbara Sanders, in contrast, found an overall effect size of 0.90 in fourteen studies of mental rotation conducted from 1975 to 1992, with no decrease over that period. Although the chronological range of their studies was considerably shorter than that of Feingold's study, Masters and Sanders did not believe that their negative results would have changed over a longer time period, since the slight change that they did observe was in the direction of *greater* sex difference.[51] A review by Larry Hedges and Amy Nowell of studies using national samples also found that sex differences have been stable over time.[52] Moreover, despite an increase in mathematics coursework taken by girls, the SAT-M gap has remained relatively stable.

Three factors limit the inferences that can be drawn about group changes in performance over time. First, the instrument used to measure performance may change, and it may be changed specifically to reduce sex differences.[53] Given the male superiority in mathematical concepts and the female in computation, for example, changes in the proportions of questions that tap these two areas would appear to reflect change in the magnitude of sex differences in mathematical ability. Second, the pool of test takers has changed over time, as an increasing proportion of high-school students take the SAT, meaning that the pool of test takers is now less "elite" than it formerly was.[54] Third, nonsignificant results are more likely to be reported today than in prior decades, so meta-analyses of more recent studies will tend to show lower effect sizes even with no underlying change in either the test or test performance.[55]

The sex differences in temperament and cognition that have been described in chapters 2 and 3 have substantial workplace implications. In part II I will discuss the effects that these differences have on how high in organizations individuals reach, what jobs they do, and how much they earn.

II Women in the Workplace

The means by which any animal "makes a living" is intimately related to the animal's physical and psychological makeup. If the physical and psychological makeup of a species varies substantially by sex, we would expect that males and females may make their livings in a somewhat different manner. The culturally universal division of labor by sex appears to be a manifestation of that principle.

Even in today's relatively egalitarian Western societies, men and women tend to seek different jobs, favor different occupational attributes, and sometimes even perform the same jobs in a somewhat different manner. Because workplace choices often influence both tangible and intangible rewards, systematically different preferences tend to result in systematically different rewards. A social environment in which individuals of both sexes are free to pursue their own priorities cannot therefore be expected, a priori, to produce identical rewards to members of the two sexes.

Sex differences in temperament and cognitive abilities, as well as occupational preferences, are at least partially responsible for a number of workplace phenomena that are sometimes labeled "problems"—the "glass ceiling," the "gender gap" in compensation, and occupational segregation. Although sex discrimination can also play a role, complete understanding of workplace patterns requires us to look honestly at other factors. Some individuals, for example, are more likely to seek, and make the requisite sacrifices and investments to achieve, the highest positions in business, government, and academia. Those who achieve positions of high status tend to be those for whom status is a high priority. Those who have high earnings tend to be those for whom high earnings are a sufficiently high priority that the sacrifices and tradeoffs necessary to achieve them are worthwhile. Because men and women vary systematically along these and other dimensions, occupational outcomes for men and women are not identical. Whether this is a problem or merely a fact is to some extent a value judgment. However, one's beliefs about the causes of the outcomes—for example, discrimination by employers or personal choice of the affected individuals—may influence the extent to which the outcomes are deemed acceptable.

In chapter 4 I will discuss reasons for the "glass ceiling"—the low representation of women at the highest organizational levels. In chapter 5 I will examine occupational segregation—the tendency of men and women to occupy different jobs. In chapter 6 I will analyze the "gender gap" in compensation—the tendency of women, on average, to earn less than men. We will see that notwithstanding arguments to the contrary, these phenomena cannot simply be ascribed to sex discrimination or pernicious socialization.

4 Once One Breaks the Glass Ceiling, Does It Still Exist?

The secret of male achievement in the world of work probably lies in the relative male insensitivity to the world of everything—and everybody—else.

—Anne Moir and David Jessel, *Brain Sex*

The relatively low representation of women at the highest level of corporate (and other) hierarchies is commonly attributed to invisible barriers that impede their progress. The glass-ceiling metaphor results in a search for causes within institutions, but a full understanding of the statistical pattern requires an understanding of the psychology of individuals as well.

Successful executives of both sexes are characterized by a constellation of stereotypically male traits such as competitiveness, assertiveness, and willingness to take risks. Indeed, willingness to take moderate risks (in both performance of the job and acceptance of assignments) appears to be a primary determinant of success. Men often view risk as opportunity, while women are more likely to view it as danger.

Women are disproportionately represented in positions from which ascension to the corporate boardroom is less likely. They are more likely than men to work in the public and nonprofit sectors, and even within the private sector they are more likely to hold staff jobs, such as public relations, than line jobs, such as running a plant.

Achievement of the highest positions typically takes decades of single-minded devotion to career, including long hours and frequent relocations. Men are, on average, more willing than women to make such an investment in part because they tend to value the fruits of that investment more. Nonetheless, most men who desire such positions, like their female counterparts, fail to obtain them. Among the self-employed, where women create their own environments rather than having to work within a corporation, even greater sex differences in hours worked and in compensation exist than are found among employed women.

The impact of families on women's representation at the highest levels is substantial, but commonly blamed factors such as inadequate day care are not likely explanations for the withdrawal of so many women from the executive track. Despite their ability to purchase child care, many find that a high-powered career is incompatible with the level of involvement with their children that they desire.

The term *glass ceiling* is a metaphor intended to describe invisible barriers to women's achievement of the highest corporate levels. It is a clever metaphor because it combines an incontestable empirical observation—the underrepresentation of women at the highest corporate levels in comparison with their overall representation in the labor force—with an assumption about its cause—powerful yet invisible forces that are holding women back and without which women would enjoy statistical parity with men at all hierarchical levels.

The glass-ceiling metaphor has been rhetorically effective in focusing attention on actions by employers and society and away from women themselves. The fact that no action by an employer can be identified as the cause of women's underrepresentation does not exonerate the employer; it simply proves the subtlety and insidiousness of the forces opposing women. If there are few women at the highest level, then there is *by definition* a glass ceiling. If there is a glass ceiling, only institutions capable of erecting it—primarily employers—can be responsible. Thus, the rhetoric encourages, if not forces, us to accept the existence of a glass ceiling from the mere existence of statistical disparities.

In fact, there is no "glass ceiling." There is an underrepresentation of women at high corporate levels compared with their representation in the work force at large. This underrepresentation has causes that need to be investigated rather than assumed, and the glass-ceiling metaphor is an impediment to understanding those causes.

The term *glass ceiling* does not, one might add, even work that well as a metaphor. A glass ceiling, after all, would not survive its first breach; it could be "shattered" only once. Yet the term is used not just to describe a level above which no women can go, but also a level above which only a few women have gone; thus, there are women both above and below it.[1] Even to adherents of the barrier metaphor, therefore, the barrier cannot be made of glass, but rather some more permeable substance. The truth is, however, that it is no barrier at all; it is merely a statistical abstraction. As we will see in this chapter, a more apt metaphor may be the "gossamer ceiling"—a barrier that women may perceive but that is not strong enough to hold back those who choose to cross it.[2]

The misleading effect of the glass-ceiling metaphor is nowhere more clearly displayed than in the work of the Federal Glass Ceiling Commission. If a watch requires a watchmaker, then a ceiling requires a ceiling maker. Thus, the commission concluded (on the basis of little evidence and after the expenditure of millions of dollars, it might be added) that the "underlying cause" of the glass ceiling is "the perception of many white males that as a group they are losing—losing the corporate game, losing control, and losing opportunity."[3] According to the commission, "white male managers view the inclusion of minorities and women in management as a direct threat to their own chances for advancement." The paradox seems to have escaped the commission: male control of the workplace is purportedly a consequence of males' perceptions that they are losing control of the workplace. Put another way, the fact that women are getting ahead is what is holding them back. Needless to say, that is not an explanation with a great deal of power; after all, what was holding women back before they were perceived to be gaining control? One seeking an understanding of the status of women in the workplace will have to look elsewhere than the Glass Ceiling Commission report to find it.

Causes of the "Glass Ceiling"

If the Glass Ceiling Commission is wrong that "white male attitudes" are an adequate explanation for the underrepresentation of women at high levels, what are the causes? One cannot rule out active present-day discrimination by employers, but neither should one assume that discrimination is all, or even most, of the story. While anecdotal accounts of women who believe that their sex has held them back get wide distribution, many other women feel that their sex has worked to their advantage.[4] There are also many men, of course, who feel, rightly or wrongly, that they were more qualified for a top position than the person who actually got it. Furthermore, when considering active discrimination, one must also take into account discrimination in favor of women. Given the intense desire on the part of many companies to promote women to the highest ranks, some part of any adverse effect of discrimination against women is surely offset by discrimination in their favor.

The "Pipeline"

Some of the causes of the glass ceiling are transitory, and the length of the management "pipeline" is an important factor. When we speak of "underrepresentation" we need to specify the relevant comparison. The most common comparison, and the one made by the Glass Ceiling Commission, is between the percentage of corporate female executives today and the percentage of women in the work force today.[5] Yet individuals who are becoming top executives today entered the management pipeline twenty to twenty-five years ago, a fact that the Glass Ceiling Report acknowledged but whose significance it apparently either failed to recognize or chose to ignore.[6] Of the fifty women profiled in a 1998 *Fortune* magazine feature on powerful businesswomen, the vast majority were in their mid-forties to mid-fifties. Thus, the relevant comparison is between the percentage of women in top positions today and the percentage entering the management pipeline in the 1970s, although as one Fortune 500 executive put it, also important is "what they have done while they are in the pipeline."[7] With the increased numbers of women who entered the pipeline during the 1980s and 1990s, one would predict a continuing increase in the proportion of women in managerial roles.

It is very easy to fall into the trap of inadvertently making inappropriate temporal comparisons. Consider the following statement from a study of women in academic psychiatry: "The proportion of women in leadership positions in academic psychiatry has not kept pace with the increase in the number of women entering the field."[8] That seems like a straightforward statement of the sort that we are accustomed to seeing; it implies that there is some "shortfall" that requires correction. But the appropriate response is, "Who would have expected otherwise?" Why would one expect that an increase in the number of women (or any other group) at the entry level would be matched in the near term by an increase in their number at the top levels?

Traits of the Successful Executive

The typical statistical comparison implicitly assumes that the temperamental and behavioral traits of high-level corporate executives are randomly distributed throughout the population, or at least randomly distributed with respect to sex. Senior executives are not a random cross-section of the population, however. They hold highly sought-after positions that require a mix of temperamental traits that not everyone possesses. Thus, the dearth of women in high places can be understood only against the backdrop of fundamental sex differences in temperament. This is not to say that biological differences are the only causes, but even without discrimination and even with such interventions as readily available day care, statistical parity between the sexes at the highest levels is exceedingly unlikely as long as individuals are free to pursue their own desires.

Successful executives have a constellation of stereotypically male traits. Both male and female executives tend to be competitive, assertive, ambitious, strongly career-oriented risk takers.[9] The personality profile of the successful executive is similar to that of the entrepreneur: "independence, desire for prestige, desire for power, internal locus-of-control belief, drive, high involvement, strong self-actualization, and moderate risk-taking."[10] The title of Michael Maccoby's study of executives—*The Gamesman*—captures an important aspect of successful executives: individuals who like to organize teams, seek challenges, and play to win.[11] Women with "masculine" personalities tend to have higher levels of career success than more-feminine women.[12] These patterns are consistent with Margaret Hennig and Anne Jardim's finding that women in managerial positions were likely to have been "tomboys" as children.[13] Successful executives tend to be willing to subordinate other things in their lives—often including their families—to maintain a single-minded focus on success, and they must make career choices along the way that will provide them preparation for leadership. Many men have some of these qualities, but few men, and even fewer women, have them all.

Now, it might be argued that this explanation confuses cause and effect. Perhaps executives have "male" temperamental traits simply as a side-effect of the fact that men occupy the positions. However, not all men possess high levels of these traits, and those who do not—like women who do not—tend not to reach the highest organizational positions. It would be quite surprising to find that rather than being competitive, assertive risk takers, successful executives were typically uncompetitive, passive, and timid.

Based upon her extensive involvement in efforts to increase the number of female executives, Felice Schwartz, the founder of Catalyst, Inc., concluded that the corporate culture's emphasis on competition and drive has made corporate environments relatively uncongenial to women.[14] Like so many others, however, she assumed that sex differences were attributable solely to socialization. As a result, she was overly optimistic in assuming that "differences in workplace behavior will continue to fade" because changes in socialization will cause "young men and women [to] grow steadily more androgynous."[15]

Leadership and Motivation to Manage

The psychological literature demonstrates a significant relationship between various measures of masculinity and femininity, dominance, and leadership perceptions.[16] Being perceived as a leader has a substantial positive impact on career achievement and allows an individual to exert greater influence. Indeed, organizational-behavior researcher Nigel Nicholson has observed that "the most important attribute for leadership is the desire to lead." While "managerial skills and competencies can be trained into a person," he says, "the passion to run an organization cannot."[17]

There is some evidence of a sex difference in motivation to manage, indicating that women are less likely than men to *want* to lead. Alice Eagly and her colleagues conducted a meta-analysis of studies measuring sex differences on the Miner Sentence Completion Scale (MSCS), which is a widely used test designed to assess an individual's motivation to manage in hierarchical organizations.[18] Overall, men scored higher than women, although the average effect size was only about 0.2. An interesting pattern emerged from the various subscales, however. The MSCS consists of seven subscales: *authority figures* (a desire to maintain positive relationships with superiors); *competitive games* (a desire to engage in competitive games and sports with peers); *competitive situations* (a desire to engage in competition with peers involving work-related activities); *assertive role* (a desire to behave in an active and assertive manner involving predominantly masculine activities); *imposing wishes* (a desire to tell others what to do and to utilize sanctions in influencing others); *standing out from the group* (a desire to assume a high-visibility position); *routine administrative functions* (a desire to meet managerial role requirements through managerial work of a day-to-day administrative nature).

Eagly and associates noted that the first and last subscales (authority figures and routine administrative functions) were consistent with female stereotypes, while the remainder were consistent with male stereotypes. They predicted, therefore, that sex differences in scores would reflect this varying congruence with stereotypes, and, indeed, that is what they found. Women outscored men on the two stereotypically female subscales, while men outscored women on the others.

The sex differences on the subscales may provide some modest insight into sex differences at various managerial levels, specifically the difference between middle management (a level at which women are abundantly represented) and top executives (where they are not). The subscales that women score higher on are traits that are important to middle management (getting along with superiors and dealing with routine administrative functions) but are less characteristic of top executives, and, indeed, may in some cases be negatively correlated with performance in, or even perhaps attainment of, top executive positions.

Sex differences in the desire to lead may result in somewhat different orientations toward the job. A woman may be more inclined to see her role as making sure that her boss looks good rather than making herself appear to be his likely successor. A 1995 survey found that while 45 percent of male executives aspired to be CEOs, only 14 percent of women did so,[19] although one cannot rule out the possibility that women's aspirations were lowered by perceptions of lesser opportunities. Moreover, three-fourths of women expressed a desire to retire before age sixty-five, compared to less than one-third of men.

The attraction of men to supervisory jobs is evident even in "traditionally female" occupations. Men are more likely to hold administrative jobs, such as social-work supervisors, head librarians, school principals, nursing directors, and presidents of nursing associations.[20] This fact is often invoked to suggest discrimination against women in promotions to supervisory positions in "helping" professions. Even in traditionally female jobs, the complaint goes, men get the "good jobs." Christine Williams has asserted, for example, that there is a "glass escalator" that whisks men to the top in these professions.[21]

Supervisory positions in the helping professions are often quite different from rank-and-file positions in ways that may differentially affect their attractiveness to the two sexes. People drawn to these professions by a desire to work with people who need help may not be motivated to seek supervisory positions that provide less of that kind of interaction. People who seek supervisory positions tend to be more motivated by a desire for money, status, and power. Given everything else we know about sex differences in the workplace, it should not surprise us to find that more of those so motivated are men. Before concluding that discrimination is the cause of statistical disparities in supervisory positions, a threshold question is whether women who want these jobs are less likely to get them than men who want them. If not, then the argument for discrimination and a "glass escalator" is weak indeed.

The Importance of Risk Taking

A taste for risk is one of the hallmarks of the successful executive.[22] A study of over five hundred top executives found that willingness to take risks was the primary determinant of success, as measured by wealth, income, position, and authority. The researchers concluded that "for most businesses, a person gets to the top by taking risks and having them work out for the best."[23] As one ascends the corporate hierarchy, the opportunity to make decisions that have an impact on corporate earnings increases. Moderately risky decisions that turn out well are the ones that most reliably enhance the bottom line. Excessive caution (as well as excessive risk taking) is generally unproductive. Significantly, female executives have been found to be more conservative in risk taking than male executives.[24] Even among professional investors, women are significantly more risk averse than men.[25]

Psychologist Elizabeth Arch has suggested that the sex differences in achievement orientation discussed in chapter 2 may be attributable in part to sex differences in risk taking. Achievement opportunities are often coupled with uncertainty and the potential for loss. Thus, they may appear threatening to women (or men) who are risk averse. Perceived risks then cause the risk averse to withdraw from achievement situations. As Arch notes, "these responses are not encouraging of participation in many of the public achievement opportunities proffered in our modern competitive, mastery-oriented cultural milieu."[26] In their book *The Managerial Woman,* Hennig and Jardim make a similar point, commenting that "men see risk as loss or gain; winning or losing; danger or opportunity," while "women see risk as entirely negative. It is loss, danger, injury, ruin, hurt."[27]

The average lesser risk taking of women can affect not only how they function in jobs but also whether they even get them in the first place. The progression to the executive ranks often requires risky career choices, and women are less likely to make such choices. Some jobs carry more career risk than others, and risk preferences influence occupational choices.[28] Line positions, such as running a plant or division, tend to carry more career risk than staff jobs, such as human resources or corporate communications. If a manager's division is losing money or his factory is putting out substandard widgets, that fact is obvious, imperiling the manager's future career progression. Although staff positions are important to the corporation, suboptimal performance in these positions is likely to be less obvious and less closely connected to the corporate bottom line. In most organizations, line positions are considered a critical part of the executive career path. A study of Fortune 1000 executives found that 82 percent identified women's lack of line experience as the most important barrier to their advancement to leadership positions.[29]

The Glass Ceiling Commission concluded that women's "clustering in relatively dead-end staff jobs" is a "barrier" to women's advancement." Yet it did not investigate whether women actually want line management jobs or whether they are willing to risk what men risk to obtain them. If women prefer staff jobs, that preference can hardly be said to be a "barrier" to achieving the less-preferred job. The message for the corporate workplace is clear, however: women (and men) who wish to progress to the top are well advised to direct their energies toward achievement of operational positions.

Different sectors of the economy are also associated with different career risk. Women managers are more likely to be found in the public or nonprofit sectors than in the private sector.[30] The Glass Ceiling Commission found that 83 percent of white and Hispanic female professionals work in the public or nonprofit sectors, compared to 56 percent of white male non-Hispanic professionals. (Presumably the commission had its own reason for the different ethnic compositions of the two groups.) Workers in these sectors have greater job security than in the private sector, but they are not likely to land in the executive suite of a Fortune 500 company.

This is not to suggest that sex differences in risk taking are the only factors leading to women's concentration in staff positions and in public sector and nonprofit jobs. Besides carrying lower career risk, these positions tend to have other attributes that may be especially attractive to women, such as lesser pressure to relocate and more-regular hours. Some of these preferences have less to do with sex differences in risk taking than they do with sex differences in family priorities.

Families and Single-Minded Career Focus

Achievement of the highest corporate positions requires more than just the right personality. It also generally requires two to three decades of devotion to one's career. This typically does not mean Monday through Friday from 9 to 5, or even Monday through Friday from 8 to 6, with the occasional Saturday morning thrown in. It means very long hours, and it often requires frequent travel and relocations to obtain broader business experience. Many women (and men) are not willing to make these investments. Even without the complication of families, women are less willing than men, apparently because the payoff—being "on top"—is not valued by women as much as it is by men. Women are more likely to say, "If this is what it takes, I don't want it." They are less willing to uproot themselves from networks of friends and relatives to move off to a new city where they do not know anyone.

Probably everyone agrees that family involvement is a major factor limiting the ascent of women in corporate hierarchies, because the presence of families exaggerates sex differences in workplace behavior. When women marry, and especially after they have children, they tend to reduce their hours and their promotion-seeking activities.[31] In contrast, when men have children, they tend to increase their work hours.[32]

Many women take extended leaves for maternity, and no matter what their intentions at the outset, many will remain out of the work force for an extended time. As Felice Schwartz observed, "The truth is that no matter how conscientious, no matter how career committed, a woman is, she can never know for certain what she'll do until she has given birth and experiences her desire to be with the baby."[33] She noted that the Liz Claiborne company found that a full one-third of its employees who went on maternity leave did not return to work. If they do return, many women cut back on their work commitment to spend more time with their children. Some reduce their hours, either remaining on full-time status or changing to part-time; some begin to decline assignments that would require travel. To an observer with an evolutionary perspective, the fact that mammalian mothers find it emotionally difficult to separate from their helpless infants is wholly unsurprising. To such an observer, it is also distressing to see women made to feel guilty for being mammals.

Many people seem to assume that what women "really want" is to be able to pursue their careers unimpeded by families and that their inability to do so requires intervention. Because everyone's choices are constrained both by societal norms and personal wherewithal, it can sometimes be difficult to discern what people's true preferences are. In examining mate preferences, we can avoid this limitation to an extent by observing people with the greatest range of choices. It turns out that high-status men and beautiful women often end up with each other.[34] So, what do couples do when both members have substantial resources, so that both members of the couple do not need to work and high-quality day care is easily affordable if they care to avail themselves of it? As psychologist Jacquelynne Eccles has noted, the general tendency for men to focus on work and women to spread their time and effort across work and family is *especially* true among those holding professional and other high-status jobs.[35] A trip to many suburban country clubs on a summer weekday would confirm Eccles's observation. Many of those present will be mothers chatting among themselves while their children are off playing in the pool. These women will be former lawyers, investment bankers, and other professionals and executives. Some will return to work when their children start school; some will not.

Why do these highly educated women leave the work force when they have their babies? Surely the absence of "quality, affordable day care" is not to blame. These women have sufficient resources to afford good day care, and many of them employ people to assist them with child care even though they are not working. The answer seems to be that they leave the work force because they want to. Although previously committed to their careers, they find that they want to spend more time with their young children than is consistent with continued employment. Whether because of their husbands' high earnings, their own savings, or inherited wealth, they have the economic resources to stay home and maintain a high standard of living, and they choose to do it.

Women with the above options are obviously the exception rather than the rule, but they are important to this discussion for two reasons. First, they reveal the preferences of many of the women with the most options. Second, these are precisely the women—talented, well educated, and well compensated—who are potential top executives and law-firm partners; that is, they are the ones whose achievements would count most toward "shattering the glass ceiling."

It is often said that male executives have an advantage in the workplace because a high proportion of their wives do not work. By marrying, says law professor Joan Williams, "elite males tap a flow of domestic services that reinforces their ability to conform to workaholic norms."[36] It is true that high-achieving men are considerably more likely to have a nonworking spouse than are high-achieving women. A study of obstetrician/gynecologists, for example, found that most of the married men were in single-earner families and that only 12 percent of the married men had wives who were professionals or executives. In contrast, 96 percent of the husbands of the married women ob/gyns were employed, and the husbands of 78 percent were professionals or executives.[37] Similarly, a study of male and female engineers found that 80 percent of the women but only 32 percent of the men had spouses who were engineers, scientists, or top professionals. Is the reason for this disparity really the "flow of domestic services" provided by the wife? It seems unlikely.

If the issue were simple economics, both members of dual-professional couples would work, and they would hire domestic help to facilitate both careers. It makes no economic sense at all for a woman earning a six-figure salary to quit her job to provide "domestic services" that could be provided by a relatively unskilled woman earning 10 percent that amount. The decision can only be understood as flowing from noneconomic motives: many women have a strong desire to spend time with their young children and have day-to-day involvement in their care, and they are willing to give up a lot of money to do so.

Other sources of information reveal that large numbers of women affirmatively desire to decrease their work-force involvement upon the birth of their children. A study by psychologist John Townsend, for example, found that only 25 percent of female medical students stated that they preferred to work full-time while their children are small.[38] A similar study among college women found that fewer than half expected to return to the work force within one year of having a baby, and of those who intended to return to work, a majority preferred to work part-time, at least until their children entered preschool.[39] As historian Elizabeth Fox-Genovese has observed, "Even highly successful women frequently want to spend much more time with their young children than the sixty-hour weeks required by the corporate fast-track will permit."[40]

Discrimination

None of the foregoing should be taken as an assertion that women face no discrimination in the workplace. However, those arguing that discrimination is *primarily* responsible for disparities between men and women often base their arguments on relatively frail evidence.

One commonly invoked line of studies are those asking subjects to evaluate a piece of written work, varying the sex of the purported author. The assertion is that these studies reveal substantial bias against women.[41] The first of these studies was conducted by Phillip Goldberg in 1968, leading to the phenomenon's being dubbed the "Goldberg effect."[42] (Interestingly, Goldberg's initial study measured evaluations by women, not by men.) Goldberg reported that higher evaluations were given when the author was identified as a man. However, the Goldberg effect is not a robust finding. Many studies either find no effect or find an effect favoring women. A meta-analysis of 106 studies that had examined the question yielded only a trivial overall effect size of -0.07.[43]

In one context of relevance here, the meta-analysis found larger effect sizes: when employment applications were being evaluated, the effect size was substantially larger ($d = -0.25$). Although an effect size of that magnitude would, by convention, be labeled "small," it is large enough to have real-world effects. However, another key finding of the study was that the effect size was substantial only when the raters had little other information to go on. When a substantial amount of individuating information was available (as is more typical in the corporate world) the effect size was negligible, which is consistent with the finding that stereotypes are most influential when there is little else to go on. That distinction does suggest that sex bias is more likely to enter into a hiring decision at the level of initial screening than in later stages during which additional information is available (such as references and interviews), and it suggests that stereotypes are more likely to bias hiring decisions than promotion decisions.

Although laboratory studies often find sex effects in simulated hiring decision,[44] studies of actual hiring and promotion decisions seem less likely to reveal such effects. Gary Powell and Anthony Butterfield examined all of the promotions to Senior Executive Service ranks (the top 1 percent of positions that are not political appointments) in a federal cabinet department from 1987 to 1992.[45] They hypothesized that bias against women would be revealed in the promotion decisions. They did indeed find that sex had an effect, but the effect was in the opposite direction from that hypothesized: the process seemed to favor women. A study based upon the National Longitudinal Survey of Youth found that in 1990, when the respondents were ages twenty-five to thirty-two, males were somewhat more likely to have been promoted (34.2 percent versus 31.0 percent), although in 1996, when they were thirty-two to thirty-nine, women were slightly more likely to have been promoted (26.0 versus 25.6 percent).[46] Much of the 1990 difference was a consequence of family effects; married men with a preschool child were most likely to be promoted, while women with a preschool child were least likely to be promoted. Indeed, never-married women were more likely to be promoted than men, whether married or not. These results hardly demonstrate rampant discrimination against women.

Current attitudes and expectations have a substantial effect on how research results are reported. For example, a recent study of hiring decisions by school principals of hypothetical administrative candidates tested the hypothesis that female principals would favor female candidates and male principals would favor male candidates. In fact, the study found that female candidates were preferred by principals of both sexes. The results are not surprising, but the interpretation the authors placed on the results was revealing. If principals had consistently favored male candidates, the results would surely have been taken as evidence of discrimination against women. The disproportionate rejection of men was given a very different spin, however, as the title of the article reveals: *Female Administrators: A Crack in the Glass Ceiling.*[47]

Some argue that even when women reach high-level positions, men do not consider them "members of the club." There is ample basis for this complaint, but it is difficult to know what to do about it. Some of the effect is at a purely social level: men are likely to talk among themselves about sports, politics, and sex, conversations that tend to exclude women, whether because of women's lesser interest in the topics, norms of propriety, or fear of sexual-harassment charges. Even more directly relevant to workplace functioning is the fact that men are more likely to be influenced by other men than by women, a pattern dating back to childhood. Especially when engaged in risky, dangerous, or otherwise consequential enterprises, men tend to form groups into which women fit precariously, if at all (a propensity of most-obvious relevance to the question of integration of women into combat units).[48] Thus, despite the fact that two of the most prestigious positions in President Clinton's cabinet—secretary of state and attorney general—were filled by women, neither woman was viewed as a particularly influential figure in the administration.

It is a comparatively easy task to require employers to promote men and women equally into leadership positions in a corporation and to pay them equally; it is far harder to ensure that they have equal influence once they are there. Leadership means more than simply occupying a "leadership position," and it entails more than objectively measurable skills. Fundamentally, leaders evoke followership, and there is an important sense—as many students of the military understand—that leaders are "born not made." Leadership training is more about giving those naturally inclined toward leadership the opportunity to grow into the role than it is about teaching leadership skills to those not temperamentally inclined to lead. Women may be less likely than men to evoke followership behavior in others, which may account for the common finding that both men and women prefer male bosses to female bosses.[49] To the extent that women's lesser influence or sense of "belonging" is a consequence of subconscious or visceral reactions of the men and women around them, it may be largely beyond the capacity of the law to deal with.

Executive Performance: Cooperation versus Competition?

That men try harder to reach the managerial ranks does not necessarily mean that they do a better job when they get there. However, the temperaments and behaviors that are influential in reaching the top ranks are not unrelated to later job performance. The literature is ambiguous on the question whether male and female managers have different management styles and personality characteristics,[50] with most finding relatively little difference.[51] The subjectivity of performance makes it difficult, however, for researchers to investigate the question in a systematic way.

It is sometimes argued that women are more inclined than men to engage in a "democratic" form of leadership under which decisions are made by group consensus rather than by fiat. This method of decision making is said to lead to better decisions because it draws on the strengths of the group's members and balances the biases of individuals; the output of the group, therefore, should exceed the sum of its parts. Although there is considerable support for the notion that women tend to favor democratic leadership styles while men favor more hierarchical ones,[52] there is little empirical support for the intuitively appealing notion that group decision making produces superior results. Indeed, if anything, the literature tends to suggest the contrary. As one researcher noted after reviewing decades of research on decision making, the literature suggests that "group performance is typically less, not more, than baseline aggregations of members' potential contributions."[53]

Relatedly, some have argued that in today's world women's more cooperative style is more productive than men's competitive style.[54] That suggestion rests on an inaccurate view of the differences between men and women, however. Although men are more inclined to competitive behavior, they are also better at shifting between competition and cooperation and cooperating with individuals they do not like, a fact that many feminists have noted through their complaints about inadequate team sports for girls.[55] Moreover, although the assumption that cooperative people will be more effective than competitive ones is reasonable, there is little evidence to support it. One study found no difference in performance between groups of highly cooperative members and groups of highly competitive members.[56] Indeed, the best results were produced by the group made up of members who were *both* highly cooperative *and* highly competitive, a temperamental pattern observed in many of the subjects. The researchers found no evidence that women performed particularly well in cooperative groups.

It is time to bury the fiction that cooperation and competition are two extremes of a single dimension, that is, that one can either be highly cooperative or highly competitive but not both. In fact, one can be high on both dimensions or on neither. As psychologist Eleanor Maccoby has noted, "among boys, cooperation and competition are by no means antithetical, but are woven into the same web of social relationships."[57] Team sports are a case in point, as the best team athletes are both very cooperative and very competitive. It is not just that they compete with their adversaries and cooperate with their teammates. Athletes often compete against members of their own team—who is going to start next Sunday?—but even that kind of internecine competition is not necessarily unhealthy.

It is likely that women perform better in some executive positions and men do better in others. These jobs are not uniform in their demands and may require a different mix of abilities, as reflected in the recognized distinction between "leaders" and "managers."[58] Some positions may primarily require an entrepreneurial spirit, while others may require "people skills." It would be consistent with much data to find that, on average, men are better in the former positions and women better in the latter.

Female Entrepreneurs

Much has been written recently about the increase in female-owned businesses, an increase that is often held up as evidence of entrepreneurial risk taking on the part of women and as evidence that the dearth of female executives in large corporations is the fault of corporations. Approximately two out of three start-up businesses in the United States are owned by women. Moreover, female-owned businesses are less likely to fail than businesses owned by men. These statistics are obviously positive signs of women's growing integration into the economy. The flip side of the low rate of business failure, however, is the tendency of businesses owned by women to be smaller and less likely to grow into large businesses.

Women often tend to pursue "lifestyle" rather than "high-growth" business ventures.[59] In 1995, only 20 percent of women-owned businesses had paid employees.[60] In 1990, about 18 percent of female-owned businesses were incorporated, compared to 31 percent for men.[61] The lesser failure rate and the lesser spectacular-success rate are probably both attributable in large part to women's lesser risk taking—a lesser likelihood of losing big generally means a lesser likelihood of winning big. Also, it should not be overlooked that contracting preferences and financial assistance available to female-owned businesses constitute a form of public subsidy for those businesses that may insulate them to an extent from economic failure.

Some of the demographic and behavioral characteristics of self-employed men and women support the existence of differences between male and female business owners. Self-employed women are more likely to be married with spouse present than are their wage-and-salary counterparts. Self-employed full-time women have been found to average 35.3 hours per week, compared to 44.6 hours for men. Among the self-employed, 71 percent of men but only 45.4 percent of women worked full-time for the full year. Self-employed full-time men work more hours on average than their wage-and-salary counterparts; women work the same irrespective of employment status. Self-employed women are about 30 percent less likely than employed counterparts to work full year, full time. While self-employed women earn substantially less than their employed counterparts (a wage ratio of 0.56 in annual earnings and 0.69 in hourly earnings), men earn more (1.14 annual, 1.07 hourly).[62]

Patterns of activity for the self-employed thus resemble those of the organizationally employed although the sex differences are magnified. Male entrepreneurs tend to be more job involved, to display more time commitment to work and less to family than women.[63] They tend to play for "bigger stakes" than women and are therefore more likely to build economically vibrant businesses but also more likely to go broke. The fact that self-employed women earn less than self-employed men even in the absence of corporations should lead one to question whether corporate misbehavior is the most likely cause of disparities among employed men and women.

Focus on the relative success of males and females should not cause us to lose sight of the fact that almost everyone who desires a top position, whether male or female, fails to get one. The progression of most men, like that of most women, stalls at some point, and this will continue to be true even in the era of "flatter" corporations. The ever-narrower organizational pyramid means that to succeed one must stand out in some way; one does not reach the top simply by keeping one's nose clean and getting one's "ticket punched." The Glass Ceiling Commission gave a sympathetic airing to a woman's complaint that ticket punching is not a guarantee of success at the highest ranks: "If I want to succeed, I have to accept the white male notion of what constitutes the good life. But even when we do that and demonstrate excellent performance by their standards, it doesn't guarantee a trip to the top."[64] She is right. It is not enough to get excellent performance evaluations; some people who do so may be good at the job they do but may not necessarily belong at higher levels. Moreover, being "excellent" on some absolute scale is not good enough; you have to be better on a relative scale than all of your "excellent" competitors. In short, the woman is correct that there is no guarantee of "a trip to the top," for her or for anyone else, and anyone who complains about the absence of a guarantee is unlikely to make it there or do a good job once there.

5 Occupational Segregation

Why Do Men Still Predominate in Scientific and Blue-Collar Jobs?

Although many previously male occupations have been integrated—indeed some have gone from being predominantly male to predominantly female—some scientific fields and many blue-collar occupations remain stubbornly male. Although discrimination is often blamed, it is usually simply assumed from the existence of the disparities. Moreover, invocation of sexism and discrimination does not explain why physics and firefighting have remained mostly male, while biology and law have not.

A full understanding of occupational patterns requires a consideration of the intersection of cognitive abilities and occupational interests. The male advantage in spatial, mathematical, and mechanical pursuits, coupled with occupational interests oriented toward "things," inclines men in different directions from women, with their advantage in some verbal pursuits and their more social orientation.

Female participation in some fields declines as they progress through the educational system. Boys and girls participate approximately equally in math and science classes through high school. With increasing educational level, female participation in fields such as physics and mathematics declines. The two most influential factors leading to the varying representation of women in science at the doctoral level are the math-intensiveness of the field and the extent of the field's social dimension, a pattern that holds even within disciplines.

Blue-collar occupations have remained even more stubbornly male than the sciences. The primary reason is the very large sex difference on the "Realistic" occupational dimension, which taps an interest in building, repairing, and working outdoors. Large sex differences in mechanical ability and physical strength also contribute to the disparity.

Many blue-collar jobs have attributes that women tend disproportionately not to like. They often have fixed hours (perhaps entailing shift work) and they tend to have worse working conditions than white-collar jobs. Many blue-collar jobs are physically dangerous, as well, and the most dangerous of them are overwhelmingly male.

Existing occupational patterns will change even if men and women do not, as job demands change over time. Strength will become less important in many jobs, and computerization of production processes may mean that more-skilled women will take the position of less-skilled men.

Despite women's entry into, and even their predominance in, many fields, others—especially scientific and blue-collar occupations—seem resistant to integration. These occupations are often labeled "traditionally male" or "nontraditional," but it is misleading to distinguish these occupations from the many in which women have become more fully assimilated on the basis of their being "traditionally male." The term "traditionally male" actually means "persistently male," since almost all occupations not specifically reserved for women were "traditionally" filled mostly by men. What distinguishes physicists and carpenters from real-estate agents and lawyers is not the traditional sex distribution of those occupations but rather the sex distributions that exist today. The interesting question is what it is about many scientific and blue-collar occupations that have caused them, unlike so many others, to remain predominantly male.

One popular explanation can be rejected at the outset. Some have argued that men have simply reserved the best jobs for themselves and left the rest for women, automatically according high status to men's jobs and low status to women's.[1] That argument, however, cannot be squared with the real world, for it fails to explain, for example, why it is that most carpenters and physicists are men but that women are abundantly represented in law and medicine. Moreover, jobs are not accorded high status simply because men hold them. As sociologist Steven Goldberg has pointed out, "it is not primarily the maleness of a role that gives the role high status, but the high status that attracts males to the role."[2] Goldberg notes that men who cannot attain high-status roles may become ditch diggers, but their maleness does not result in ditch digging becoming a high-status occupation.

The correct observation that the highest-status jobs tend to be occupied disproportionately by men and the lowest-paying jobs tend to be occupied by women is often transformed into the incorrect belief that men get all the "good" jobs and women get all the "bad." Obscured in this discussion is the fact that men's jobs are often less attractive than women's. Despite the male predominance in the highest-status jobs, jobs held by women are rated overall as being as high or slightly higher in status than jobs held by men.[3] Consistent with greater male variability on a variety of measures, men tend not only to hold the highest-status jobs but also the lowest-status ones. Warren Farrell has pointed out that men have a virtual monopoly on the least attractive jobs.[4] When jobs are rated on a combination of salary, stress, work environment, outlook, security, and physical demands, twenty-four of the twenty-five "worst" jobs were overwhelmingly male. Among the twenty-five worst jobs were the blue-collar occupations of seaman, cowboy, roustabout, construction laborer, police officer, truck driver, fisherman, and farmer.[5]

An explanation more nuanced than male oppression is therefore needed to account for women's complete integration in some fields and the persistent sex segregation in others. Part of the explanation lies in the sex differences in cognitive abilities discussed in chapter 3. Occupations have dramatically different ability requirements. Verbal and linguistic skills are most important for careers in the humanities, for example, while quantitative and spatial abilities are most important to careers in math and science. Although some of the cognitive sex differences are quite large and could by themselves account for a substantial portion of sex disparities in occupations, they are not large enough to explain the entire difference.

Interests and values are as important as cognitive ability in occupational choice. A person with abilities that qualify him for a job that is inconsistent with his interests and values will probably not select that job, or, if he does, he is likely to be unhappy in it. As psychologist David Lubinski has put it, vocational psychologists ask clients, in one form or another, two questions: "Are you able to do it? Are you happy doing it?"[6]

The questions posed by Lubinski have been somewhat formalized into the "Theory of Work Adjustment,"[7] according to which two dimensions of correspondence between the individual and the job are required for a successful match, *satisfactoriness* and *satisfaction*. Satisfactoriness involves correspondence of the individual's abilities and the demands of the occupation, while satisfaction entails correspondence of the occupational rewards (such as compensation, working conditions, and type of work) and the individual's values and interests. These dimensions reflect the common-sense proposition that people gravitate toward, and do best in, jobs for which they have the skills and ability and that provide them the kinds of satisfactions they desire from a job.

The extent of correspondence on both measures has a substantial effect on the work experience. Individuals who "fit" the organization will be more likely to be attracted to it in the first place, and, if they secure employment, they are more likely to be favorably evaluated by the organization and to exhibit greater motivation and better performance than those who do not.[8] The importance of fit suggests caution in encouraging people to pursue occupations for which they may have the talent but lack the interest, even if the occupation is a highly paid and prestigious one.[9]

In this chapter I will focus on two areas of occupational segregation by sex that have garnered the lion's share of attention—women in mathematics, science, and technology and women in blue-collar occupations. We have already discussed sex differences in cognitive abilities—an important component of "satisfactoriness"—but before turning to these occupations, we should briefly examine sex differences in occupational interests, which are important to "satisfaction."

Sex Differences in Occupational Interests

Measurement of occupational interests has become quite sophisticated, and vocational counseling relies heavily on these measurements. Two widely used instruments are the Strong Interest Inventory and the Self-Directed Search. Both rely heavily on Holland's "hexagon" of interests, which captures aspects of personality. The six General Occupational Themes are represented by the acronym RIASEC: *Realistic* (people who enjoy building, working outdoors, and working with things rather than ideas or people), *Investigative* (people who are interested in analyzing abstract problems and understanding the physical world), *Artistic* (people who enjoy self-expression, creating or enjoying art, drama, music, and writing), *Social* (people who are sociable, helping, instructing, and humanistic), *Enterprising* (people who enjoy persuasion and leadership and are impatient with precise work and sustained effort), and *Conventional* (people who prefer highly ordered activities, such as accounting, organizing, and processing data). Holland types that are close on the hexagon, such as Realistic and Investigative, are more similar than those that are opposite, such as Realistic and Social.

The Holland types are "ideal types," and individuals are blends of varying amounts of the different themes, resembling some types more than others.[10] A particular individual might, for example, resemble the Investigative type most strongly and then, in declining order, the Realistic, Conventional, Enterprising, Social, and Artistic. In practice, only the three dominant categories are usually recorded, so our hypothetical individual would be an "IRC." Occupations are categorized by three-letter codes, and the occupations that are coded "IRC" include nuclear-fuels research engineer, computer programmer, material scheduler, toxicologist, and pulmonary-function technician.[11] Had the order of the six categories been exactly reversed, the individual would have been an ASE. Occupations coded ASE include composer, story editor, playwright, dancing instructor, public relations representative, and art teacher. Probably few IRC individuals would be inclined to consider ASE occupations, and vice versa.

The Strong Interest Inventory has two other components relevant to our discussion. In addition to the six General Occupational Themes, there are twenty-five Basic Interest Scales, and, newly added in the 1994 revision, four Personal Styles.[12] The Basic Interest Scales are subdivisions of the broader General Occupational Themes.[13] They tap specific areas of interest such as office work, mechanical activities, science, social service, sales, and so forth. The Personal Styles are Work Style (preference for working alone with ideas, data, or things versus working with and helping others), Learning Environment (preference for practical versus academic learning environments), Leadership Style (preference for doing things oneself versus directing the work of others), and Risk Taking/Adventure (preference for risk taking and adventure versus caution and safety).

Substantial sex differences are found in all three components of the Strong. A large-scale study found significant sex differences for all six General Occupational Themes, with effect sizes ranging from quite large to quite small: Realistic (1.28), Investigative (0.56), Artistic (−0.29), Social (−0.29), Enterprising (0.19), and Conventional (0.06).[14] These results are typical; males reliably score higher on Realistic, Investigative, and Enterprising, and females score higher on Artistic and Social. Sex differences in the Conventional theme are small enough that they are often not statistically significant unless the sample size is large.[15]

Sex differences appear on a number of the Basic Interest Scales, as well. Kaufman and McLean found sex differences with effect sizes (absolute value) of 0.40 or greater on Domestic Activities (−1.06), Office Practices (−0.83), Social Service (−0.62), Art (−0.55), Music/Dramatics (−0.40), Adventure (1.21), Mechanical Activities (1.29), Athletics (0.81), Military Activities (0.66), Science (0.65), Mathematics (0.53), Agriculture (0.49), and Law and Politics (0.49).

Two of the Personal Styles also show marked sex differences. The first is Work Style, which measures whether people prefer to work with people or with ideas, data, or things. The Work Style indicator derives from the work of Dale Prediger, who suggested that the Holland hexagon can be summarized along two dimensions, a "data-ideas" dimension and a "people-things" dimension. While sex differences are not typically found on the "data-ideas" dimension, substantial sex differences are reliably found on the "people-things" dimension, with differences of a full standard deviation or more.[16] The latter finding should not be surprising given the tendency of girls, from infancy, to be more people oriented and boys more thing oriented. Of 109 occupations examined in the norming of the Strong, the 5 occupations at the people end of the Work Style scale were child-care worker, home economics teacher, community service organization director, and secretary, while the 5 at the low end were physicist, chemist, mathematician, computer programmer/systems analyst, and biologist.[17]

The other Personal Style of interest here is the Risk Taking/Adventure style. This style largely replicates the Adventure Basic Interest Scale from earlier versions of the Strong, and scores tend to decrease substantially with advancing age.[18] The sex difference on the Adventure scale was reliably one of the two largest on the Basic Interest Scales ($d = 1.21$ in the Kaufman and McLean study), the other being mechanical activity ($d = 1.29$). The highest scorers on the Adventure Personal Style are police officers, whereas the lowest are dental assistants. The Adventure scale is highly correlated with the personality trait of sensation seeking, and it identifies individuals who are interested in taking both physical and social risks.[19] Women with high Adventure scores tend to gravitate to nontraditional occupations, as well as marrying later and desiring fewer children.

Given the dramatic changes in the position of women in the workplace in the last several decades, one might have expected that sex differences in occupational interests would have exhibited correspondingly dramatic decreases, but they have not. Reviewing fifty years of results on the Strong Interest Inventory and its predecessors, Jo-Ida Hansen, although noting some changes, characterized the study as indicating "tremendous stability" of male and female occupational interests and the differences between them.[20]

The vocational-interest literature demonstrates that exclusive focus on an individual's cognitive abilities is unlikely to yield accurate predictions of occupational interest or success. For example, individuals assigned Holland three-digit codes of SIA (characteristic of clinical psychologists and general-duty nurses) are unlikely to seek out and prosper in a field like physics (an IRE occupation) even if they possess the requisite level of mathematical ability. People with high Social codes tend not to thrive in the often-cloistered environment of laboratory science, while those entering math-intensive fields tend to have a "low need for people contact."[21]

Women in Science and Technology

Increasing the representation of women in scientific and technological fields is one of the leading priorities of many who believe that justice requires society-wide integration. The sociological explanations for the dearth of women in science are familiar: girls are "channeled" away from math and science by parents and teachers and taught that math and science are for boys. Rather than taking science and math in high school, they are directed into other, more "feminine," fields. Those females with the temerity to enter this "well-fortified bastion of sexism" are driven out by the entrenched antifemale hostility. This perceived hostility is so great that philosopher Sandra Harding has pronounced it "shocking . . . that there are any women in science at all."[22]

The sociological explanation raises the initial question of just why science is more subject to sexist resistance than other fields. The medical profession formerly had the image of being comprised of (mostly) men who thought they were "gods," and the legal profession was full of (mostly) men who viewed themselves as knights errant roaming the corridors of power doing justice and exerting their will. Professions made up of omnipotent gods and chivalrous knights might have been expected to be the last bastions of sexism, but these citadels crumbled quickly before the onslaught of the female hordes. Over 40 percent of new doctors and lawyers are women. Who, then, can stand before the onrushing tide of female power? The surprising answer turns out to be the mathematics and engineering geeks, fighting to keep pocket protectors out of the hands of the fair sex.

The Sex Disparity in Scientific Fields

At the doctoral level, which is primarily where those critical of low female science participation focus, there is a wide variation by field in female representation. In 1995, women earned 12 percent of the doctorates in engineering,

12 percent in physics, 31 percent in chemistry, 41 percent in biology, and 64 percent in psychology. Within fields, there is substantial differentiation among subfields. Women were sparsely represented among mining/mineral and mechanical engineering (0 percent and 6 percent, respectively), but more heavily represented in environmental health and bioengineering (25 percent for both). In physics, women received 17 percent of the doctorates in acoustics but only 4 percent of the doctorates in elementary particle physics. In biology, women earned 22 percent of the entomology degrees, but 68 percent of those in nutritional sciences. In psychology, women earned 38 percent of the degrees in physiological psychology but 80 percent of those in developmental and child psychology. In the social sciences, women were lightly represented in statistics and economics (23 percent and 24 percent), and heavily represented in anthropology and sociology (58 percent and 53 percent).[23] Often neglected in discussions of women in science are women in applied-science professions, such as medicine and veterinary science. Women make up more than 40 percent of new doctors and approximately 70 percent of new veterinarians.[24]

Some people are troubled by this pattern. The director of women's programs at the National Science Foundation (NSF) has characterized the fields in which there has been little increase in female participation, such as physics and engineering, as "the ones where the stereotypes have been preserved."[25] This argument is wholly circular, of course: women are underrepresented because there is a stereotype that these are men's jobs. How do we know there is such a stereotype? Because women are underrepresented. However, causation is something that must be proved, not assumed. There is a stereotype that basketball players are tall, and if we go to an NBA game that stereotype will be reinforced. That hardly proves, however, that tall people become basketball players and short people do not *because of the stereotype.*

If hostility toward women accounts for the pattern of female participation, it is a quirky and selective hostility. A more plausible explanation is differential interest and ability, just as the Theory of Work Adjustment would predict. The fields and subfields avoided by women are those that tend to have the lowest social dimension—mechanical engineering, particle physics, mathematics, entomology, physiological psychology, statistics, and economics—while those attracting relatively large numbers of women—such as anthropology, sociology, biology, developmental and child psychology, nutritional sciences, environmental health, and bioengineering—have a higher social dimension. Moreover, the fields avoided by women tend also to be the most mathematically demanding. Given the relative positions of males and females on the "people-things" dimension and the disproportion of men at the highest levels of mathematical ability, it would be astonishing to see the sexes sort themselves uniformly throughout these fields.

Those who believe that women's relatively slow advancement in scientific occupations is a consequence of male resistance to women are faced with a paradox: women seem to have made the least progress in occupations that tend to provide the most concrete measures of successful job performance. As psychologist Doreen Kimura has commented: "Why anyone should imagine that [a conspiracy against women] could be maintained in a manifestly egalitarian discipline is never made clear. Science, more than most disciplines, has quite explicit rules of evidence and fairly objective criteria for excellence. We might therefore expect success in science to be, if anything, *more* rather than less related to merit, than in other areas of scholarship."[26] Instead, however, it is in academic disciplines in which assessment of scholarly quality is often very subjective (the humanities and social sciences as opposed to the physical sciences and engineering) that women have made the greater strides.

In trying to understand the pattern of female representation in science, we need to start back in high school, for it is in high school where girls are supposedly sapped of self-confidence and where they are discouraged from participation in math and science.

Girls and Boys in High School

A brief look at the experience of girls in high school shows the weakness of the argument that they are channeled away from math and science courses. At every level of math before calculus, girls equal or outnumber boys. Only at the level of calculus does the proportion of boys slightly exceed the proportion of girls (11.2 percent versus 10.6 percent in 1998). In the sciences, the story is much the same. Girls are slightly more likely than boys to take biology (including AP/honors biology) and chemistry, and boys are slightly more likely to take physics. On the other hand, boys are more likely to drop out of high school and more likely to repeat a grade than girls.[27]

Girls' performance, as well as participation, is roughly equivalent to that of boys. Girls consistently receive higher grades in all subjects, including math and science, in high school. On the mathematics portion of the National Assessment of Educational Progress (NAEP), which is taken by students at all ability levels, male and female twelfth graders perform equally well on average, although boys are somewhat more likely to score at the "advanced" (3 percent versus 1 percent) or "proficient" (18 percent versus 14 percent) levels. On the science portion of the NAEP, boys score only slightly higher than girls. Nonetheless, boys outperform girls on the SAT-M, which is taken by a more elite sample of students, exceeding girls' scores by approximately 35 points both among all test takers and among those who have taken calculus and physics.[28]

What about differential encouragement of boys and girls? We hear anecdotes of girls being discouraged from pursuing math and science by guidance counselors, teachers, and parents, who tell them not to worry their pretty little heads thinking about becoming physicists or doctors, and who instead urge them to become nurses and teachers. There is little empirical evidence that such discouragement is widespread, however, and, given the self-conscious attempts of schools to encourage girls in science, there is much reason to think that it is not. A recent study of adolescents' career interests in science found that girls were less likely than boys to believe that they had not received serious attention from teachers about science.[29] Similarly, college girls are more likely than college boys to report that they chose science majors because of encouragement from parents or teachers, while boys are more likely to report that they chose science because of long-term interest in the subject.[30]

Looking toward college, the picture changes a bit. Girls take fewer of the Advanced Placement (AP) exams that allow high-scoring students to earn college credit. In 1996, girls constituted 47 percent of basic calculus AP candidates but only 38 percent of the takers of the more advanced calculus exam. Similarly, girls are 20 percent of takers of the basic computer science AP exam, but only 12 percent of takers of the more advanced test, and in physics, girls are 35 percent of takers of the basic physics AP test, but only 26 percent and 22 percent of takers of the two more advanced exams were girls.[31]

What Happens in College?

If boys and girls have roughly the same experiences with math and science in high school, why do fewer girls pursue math and science in college? Part of the answer is that girls' aversion to college math and science is not as great as many people believe and is not uniform among scientific fields. In 1996, one-third or more of bachelor's degrees in astronomy, earth sciences, and oceanography, and over 40 percent of degrees in chemistry and in mathematics and statistics were awarded to women, as were a majority of degrees in biological sciences and a whopping 73 percent of psychology degrees. On the other hand, only about 18 percent of bachelor's degrees in engineering and in physics went to women.[32] Although there has been a dramatic increase in female participation in many of these fields in the last thirty years, readers may be surprised to find that even in 1966, women earned about one-third of the bachelor's degrees in mathematics and computer science.[33]

Part of the sex difference in mathematics and science participation reflects the increasing sexual disparity in mathematical talent at more rarified levels. We saw in chapter 3 that substantially more males than females occupy the very highest levels of mathematical ability. That fact alone would suggest a sexual disparity in math-intensive occupational fields. For takers of the Medical College Admission Test, for example, among the top 3.7 percent of scorers on the physical sciences subtest, males outnumber females by approximately 4 to 1.[34] Just as important, however, are differences *among* the mathematically talented. Although we tend to view the "gifted"—often defined as the top 1 percent of mathematical talent—as a relatively homogeneous group, in fact they are highly diverse. The range of the top 1 percent of most ability groups is as broad as the range from the bottom 2 percent to the top 2 percent. In a typical IQ test, for example, which is standardized with a mean of 100 and a standard deviation of 15, the range of the top 1 percent runs from about 135 to over 200, which is as broad as the range from 66 to 134.

Males especially outnumber females in the top quarter of the top 1 percent of mathematical ability, and while it might seem that there is a point of diminishing returns beyond which additional ability has no payoff, that does not seem to be the case. On a variety of criteria, such as earning a degree in science, level of college attended, grade-point average, and intensity of involvement in math and science, significant differences exist between the top and bottom quarters of the top 1 percent.[35]

Even among the mathematically gifted, there is a large sex difference in the extent to which college students exhibit a single-minded dedication to math and science. Lubinski and Benbow reported that during a four-year period at one university, gifted females enrolled in math and science courses and English and foreign language courses in approximately equal proportions, while males were six times as likely to enroll in math and science courses as in English and foreign language.[36] These women were following a different path from the men not out of an aversion to math and science or a lack of confidence in their abilities, but because they "are more socially and esthetically oriented and have interests that are more evenly divided among investigative, social, and artistic pursuits."[37] This difference is reflected in the decision to advance to higher levels of education, as well. Lubinski and Benbow found that approximately 8 percent of mathematically gifted males, but only 1 percent of gifted females, were pursuing doctorates in mathematics, engineering, or physical science.[38]

High mathematical and verbal abilities are generally sufficient for entrance into scientific fields; spatial and mechanical ability are not explicitly screened for. It appears, however, that individuals self-select into certain occupations—such as engineering and the physical sciences—based also on their spatial and mechanical ability, a process that would tend to exaggerate sex differences in representation even more.[39] Indeed, there is so little screening for spatial and mechanical ability—and so much for mathematical and especially verbal abilities—that many spatially talented students who would have both the ability and interest to pursue technical professions may be overlooked by the educational system.[40]

Persistence

Many have expressed concern that the math and science pipeline is "leaking" all along its length,[41] raising the question of just why it is that more girls than boys turn away from science. One important reason is that girls are more likely than boys to find science "boring," an attitude that apparently continues into adulthood. Surveys have found that women at all ages from twenty to seventy-nine have less confidence in science than males and are more likely than males to agree with such statements as "science breaks down people's ideas of right and wrong" and "science pries into things."[42] A study of science attitudes in adolescents found that girls were less likely than boys to view science as a "fun puzzle to study." It also found another interesting sex difference. Girls were substantially more likely to participate in science fairs than boys but less likely than boys to engage in science activity at home.[43] This difference is similar to patterns in other areas. For example, girls are now quite likely to play on organized sports teams, but not very likely to participate in spontaneously organized team sports.[44]

The importance of ability to persistence in science is often underestimated. When studies control for such measures of ability as entrance examination scores or undergraduate science grades, the sex disparity diminishes or disappears.[45] A study of science persistence in four elite undergraduate institutions found that, controlling for performance in basic science courses, women were no more likely to leave engineering, biology, chemistry, or physics than men. They were, however, more likely to leave mathematics and computer science.[46] (Similarly, at the Ph.D. level, women fail to complete their Ph.D.'s at a substantially higher rate than men, but controlling for ability eliminates most of the sex difference.)[47] Women in the undergraduate study were also more likely to complain that science courses are too competitive and to have less confidence in their abilities, even controlling for performance. They were also somewhat less likely to do unassigned reading in scientific fields than men. One interesting finding of the study related to the issue of whether there is a sex difference in college-level science grades. Science departments often offer courses specifically for nonscience students, which tend to have higher curves. This study found no sex difference in grades in science courses for nonscience students, but a significant sex difference favoring males (grade-point average of 2.98 versus 2.75) in standard introductory science courses.

One important sex difference that may affect women's persistence in science is that, as we saw in chapter 2, females seem to need more positive feedback and more nurturing environments in order to maintain high motivation.[48] One of the difficult transitions from high school to university for many top students, especially women, is the shift from an environment in which they are stars lavished with attention to an environment in which they are relatively anonymous and surrounded by many other people as bright as they are. While their high-school science teachers knew them well and encouraged them to continue in science, in the university they are largely anonymous to their professors, especially as they take their introductory courses.[49] In their study of attrition of women from scientific fields, sociologists Elaine Seymour and Nancy Hewitt found that women especially disliked the large introductory classes because they felt the professors did not care about them.[50] Although faculty tended to interact with male and female students in the same impersonal way, the contrast from high school was greater for the females because of the extra personalized attention that they had previously received.

International Comparisons

The low representation of women in the physical sciences is not uniform across cultures. Although in 1994 women represented less than 5 percent of physicists in the United States, Britain, and Canada, women represented 30 to 47 percent of physics faculty in Hungary, Portugal, and the Philippines.[51] At one of Turkey's best universities, one-third of physicists and mathematicians are women, as are two-thirds of the chemists.[52] These figures are often cited to suggest that barriers to women are purely cultural, but that argument is persuasive only if universities in these countries were shown to attract the most mathematically and scientifically able individuals, as they do in the United States. If, instead, highly talented individuals gravitate toward industry rather than academia because of low academic salaries, as seems to be the case in many of the countries,[53] or if they emigrate to countries with more advanced scientific establishments, then the distribution of the sexes in the universities does not tell us very much. Perhaps it is significant that the percentage of women in the hard sciences seems to be *negatively* correlated with the status of women in the society.

No one contends that women have not faced obstacles in their move to integrate the scientific profession. Surely they have. But the most relevant question is "why science?" Women have seemingly been unable to overcome obstacles in science that they quickly overleapt in other disciplines. Although it has become usual to ascribe women's difficulty to the "male culture" of science, a century ago the cultures of biology and law were scarcely less male than that of physics. Cultural pressures alone are simply inadequate to explain the observed patterns.

Blue-Collar Jobs

Sexual integration of blue-collar jobs has been less complete than integration of white-collar jobs, including management and the professions. From 1983 to 1997, women's representation in blue-collar occupations remained stable at around 18 percent.[54] Unlike white-collar occupations, however, blue-collar occupations have drawn comparatively little attention from either activists or scholars. Two decades ago, Kay Deaux and Joseph Ullman observed that "behavioral research on blue-collar workers is less fashionable than the study of white-collar workers."[55] This bias continues to this day, probably in substantial part because the white-collar investigators of occupational sex differences view white-collar jobs as both more important and more interesting than blue-collar jobs. The plateauing, and even the decline, of wages in many blue-collar occupations may also make their integration seem less urgent. Nonetheless, many have advocated greater sexual integration into skilled and semiskilled blue-collar occupations.

Sexist Stereotyping and Harassment

The conventional explanation for the low representation of women in blue-collar occupations focuses on sexist stereotypes of appropriate work for women and the sexual harassment that women sometimes face when they take such jobs.[56] Both of these factors do indeed play a role, and though these factors are also invoked to explain the low representation of women in science, they may have greater explanatory power in the blue-collar context. One reason that stereotypes may operate more strongly on blue-collar occupations is that schools, which today self-consciously encourage girls to participate in math and science, provide little training for blue-collar occupations and therefore have little occasion to encourage girls to pursue them.

Women's disinclination to pursue blue-collar occupations is not necessarily just a consequence of lack of exposure. A study examining the interest in blue-collar work among white-collar female employees of a utility company found that women who had performed blue-collar work for the utility during a strike were *less* inclined to consider blue-collar occupations than women who had not.[57]

Women do, of course, sometimes face discrimination in blue-collar jobs. Often, however, the special treatment that they receive is positive, such as relief from particularly burdensome duties.[58] While one might decry this special treatment as paternalism, women who receive this special treatment often appreciate it.[59]

As for harassment, women are, for a variety of reasons, more likely to face serious forms of harassment in blue-collar occupations than in white-collar ones.[60] Class norms may make harassment more acceptable in blue-collar occupations, and the physical orientation of many blue-collar occupations may result in conflict being manifested in physical ways. Moreover, men in blue-collar occupations (especially dangerous or physically demanding ones) often believe that women are less capable than men and are therefore likely to pose a danger to them or require them to work harder to make up for women's lesser physical strength. For example, the negative attitude of male police officers toward female officers stems from their concern that they will not be able to depend on the women in those relatively infrequent circumstances when confrontations turn violent,[61] a concern also voiced by men in the military.[62] Others have suggested that blue-collar men are especially dependent upon their jobs for their masculine self-image, and that self-image is threatened by the integration of women. Driving women out of the workplace, under this reasoning, results from a desire to protect that masculine self-image. In contrast, sociologists Irene Padavic and Barbara Reskin, although not doubting the reality of sexual harassment, have argued that negative attitudes of co-workers have been exaggerated as a reason for women's lack of attraction to blue-collar work.[63]

Factors Depressing the Representation of Women in Blue-Collar Occupations

Stereotypes, harassment, and unwelcoming environments are not the whole story and are probably not even the majority of the story. Indeed, female blue-collar workers report job satisfaction as high as, or higher than, that of male blue-collar workers.[64] Rather, the low representation of women in most blue-collar occupations is due in substantial part to a variety of factors that we have already examined. Sex differences that appear to exert an important influence on the representation of women in blue-collar jobs include differences in occupational interests (particularly in the Realistic dimension of the Holland hexagon), abilities (primarily mechanical ability and physical strength), and in job-attribute preferences, such as a disinclination among women toward physically strenuous, dirty, and dangerous occupations and a preference for flexible hours.

The Holland Realistic Dimension. Most blue-collar occupations are heavily oriented toward the Realistic dimension of the Holland hexagon, which, it will be recalled, taps an interest in building, repairing, and working outdoors, and a preference for working with things rather than people. In fact, the three-letter Holland code for virtually all blue-collar jobs begins with R. Examples include arc welder (RIS); electrician (RIE); laboratory tester (RIC); ambulance driver (RSI); butcher (RSE); press operator (RSC); plumber (construction) (REI); firefighter (RES); locksmith (REC); automobile mechanic (RCI); roustabout (RCS); and tractor operator (RCE).[65]

The Realistic dimension exhibits the largest sex difference of the Holland themes—with the effect size often exceeding 1.2. Consequently, knowing nothing about the sexes other than their typical Holland profiles would lead to a prediction that men would predominate in blue-collar jobs. If the scores are normally distributed with equal variances, an effect size of 1.2 would mean that females would constitute less than one fifth of those with Realistic scores at the male mean and less than one-ninth of those in the top quartile of the male distribution. Thus, controlling for just a single Holland dimension may account for the bulk of the sexual disparity in blue-collar occupations.

Abilities. Sex differences in abilities are also important to the distribution of the sexes in blue-collar jobs. Many blue-collar occupations require mechanical and technical ability, many require substantial physical strength, and some require both.[66] For example, heavy-equipment mechanics must have a high degree of mechanical ability, but many of the tasks they are called upon to perform require substantial upper-body strength as well.

Very large sex differences exist in mechanical interest ($d \approx 1.3$ on the Basic Interest Scale) and mechanical ability ($d \approx 0.9$ on the DAT and AFOQT). Given the undoubted substantial correlation between mechanical interest and ability, on the one hand, and a Realistic orientation on the other, these dimensions are clearly not fully independent. Nonetheless, to the extent that the correlation between the two is less than perfect, they would have a cumulative effect on the expected sex ratio.

Substantial sex differences exist in physical strength, as well. Women have approximately one-half to two-thirds the upper-body strength of men.[67] In many studies, the effect sizes are greater than 1.5, which means that there is very little overlap between the strength distributions of the two sexes.[68] For most measures of upper-body strength the probability that a man selected at random from the population will be stronger than a randomly selected woman is over 95 percent. These differences substantially exceed in magnitude sex differences in psychological traits.

Job Attributes. In the next chapter we will discuss in more detail sex differences in job-attribute preferences. For now, suffice it to say that attributes of white-collar jobs—flexible hours, safe and clean working environments, and social contact—tend to be more consistent with women's preferences than are attributes of blue-collar jobs. Many blue-collar jobs entail shift work, which, although not favored by either sex, is viewed especially negatively by women.[69] While the most obvious reason for women's allergy to night shifts is that it takes them away from their children, the strong preference that women have for day work is not limited to mothers.[70]

Blue-collar occupations tend to have worse working conditions than white-collar occupations. Many involve work outside, which in some cases can be pleasant, of course, but extremes of heat and cold coupled with rain and snow can often make work outdoors rather unpleasant. Yet other jobs entail unpleasant conditions inside, whether the heat of a smelter, the noise of an industrial plant, the darkness of a coal mine, or the smells of a chemical plant.

Physical Danger. One strong preference that women have is for jobs that are physically safe, and many blue-collar jobs are quite dangerous. The most dangerous of the blue-collar jobs tend to be jobs that are overwhelmingly male. In 1995, for example, the ten occupations with the highest risk of death (with the relative risk compared to the average worker in parentheses) were fisherman (21.3); logger (20.6); airplane pilot (19.9); structural metalworker (13.1); taxicab driver (9.5); construction laborer (8.1); roofer (5.9); electric power installer (5.7); truck driver (5.3); and farm occupations (5.1). Thus, fisherman was the most dangerous occupation, with a risk of death over twenty-one times the fatality rate for the labor force at large. The causes of death in dangerous occupations are largely, though not entirely, what would be expected: drowning for fishermen, plane crashes for pilots, motor vehicle accidents for truck drivers, electrocution for electricians, and falls for roofers and metalworkers. Surprisingly, in light of how they drive, only about one in five deaths of taxi drivers results from a motor-vehicle accident; instead, the substantial majority are homicides in the course of robberies.[71] With the possible exception of the broad and somewhat amorphous "farm occupations" category, all of the above occupations are less than 10 percent female and most are less than 5 percent.

Interaction of Multiple Traits Exaggerates Disparities in Outcome

People often respond to suggestions that biological and psychological sex differences account for sex differences in occupational outcomes by focusing on a single factor. Sex differences in mathematical ability are not enough to explain some outcome, they may say, or sex differences in physical strength are not enough to explain another. Usually that objection is technically correct, but seldom is one factor alone responsible. Several factors may interact to produce a sex disparity far greater than might be predicted from a single factor.

Consider the case of firefighters. Much concern has been expressed over the low number of women in firefighting. As of 2000, only 36 of 11,000 firefighters in the New York City Fire Department were women,[72] and only a dozen out of 1,600 firefighters in Boston were women.[73] Many find these numbers troubling, believing that recognized sex differences are not large enough to explain these numbers and that therefore some invidious force must be at work.

It is possible that there are fewer female firefighters than one would expect based solely upon individual interests and capacities, but not dramatically fewer. A number of attributes of the occupation preclude a rational expectation of anything like parity in a major metropolitan fire department. The Holland occupational code for firefighter (RES) ($d \approx 1.2$ for the Realistic dimension) would by itself predict a large sex disparity. The combination of the Holland code and sex differences in the Strong Interest Inventory Risk Taking/Adventure Style ($d \approx 1.2$) would predict a very large sexual disparity. Yet these sex differences are smaller than yet another sex difference, the difference in physical strength. Unlike many jobs in the information and machine age, the job of firefighter continues to demand a substantial measure of physical strength, and the sex difference ($d \approx 1.5$) in the upper-body strength needed to manipulate a fire hose, set up a twenty-foot ladder, or rescue a fire victim is about as large a sex difference as one is likely to find. Based upon the foregoing effect sizes, if the job of firefighter requires a Holland Realistic score, an Adventure score, and physical strength all in the top quarter of men, then (assuming that these traits are independent, admittedly a big assumption) one would expect a sex ratio of more than 1,000 men for every woman. In addition, there are many other features of the firefighter job that are disproportionately unattractive to women, such as long shifts necessitating sleeping over in the firehouse, which would further increase the sex ratio. While one might fairly quibble with the particular numbers chosen for this example, any way one slices it, there will be a very large disparity.

The actual experience of the New York Fire Department is roughly similar to what would be predicted by the above numbers. In order to obtain a position in the department, an applicant must pass both written and physical tests. In a recent application cycle, approximately 850 women registered to take the written test, but about half dropped out of the process before taking it. Only 450 (2.6 percent) of the 17,000 takers of the written test were women. Of those, 354 passed, but only 105 followed through by appearing for the physical test. That test consists of eight activities that simulate actual job tasks, such as raising a twenty-foot ladder and manipulating a water-filled fire hose, with each part being scored on a pass-fail basis. Because large numbers of men usually pass all eight parts, passage of all eight has become virtually a de facto requirement for hire. Of the 105 women who took the physical test in the recent cycle, 11 scored high enough on the physical test to be hired, including one who had a less-than-perfect score but who was entitled to preference under a program for emergency medical technicians. In contrast, 57 percent of the men who took the physical test registered perfect scores.[74]

Implications for the Future

Existing patterns do not necessarily suggest that the work that men tend disproportionately to do will continue to be done by men in the same proportions. Even if women's job preferences were not to change, jobs do. Tasks that required physical strength last year may be performed by high-tech equipment next year; work that is directed on a factory floor today may be directed by computer from an air-conditioned office tomorrow. Economist Bruce Weinberg has found that over half of the increase in demand for female workers is attributable to an increase in computer use. Increasing computerization of production processes means that more highly skilled women can replace less-skilled men who prior to computerization had a comparative advantage because of the difficult physical requirements and unpleasant working conditions associated with the work.[75]

To say that women and men will, by capacity and inclination, decline to gravitate to blue-collar and scientific occupations in the same proportions is not to suggest that the precise distribution that prevails in occupations today is in any sense the "correct" one—that is, the equilibrium distribution that would prevail in the absence of all arbitrary barriers to individuals' pursuit of their own preferences. Indeed, the constantly changing profiles of both occupations and workers suggests that there is probably no long-term equilibrium at all. However, for reasons set forth in this chapter, it is extremely improbable that the "correct" distribution in the fields that have remained persistently male is a 50:50 sex ratio; indeed, the "correct" distribution is probably much closer to what prevails today than to sexual parity. While we should be alert to artificial barriers to women's pursuit of predominantly male occupations, we should not assume their existence simply from the lack of parity. A 50:50 representation of males and females in all occupations would be far more indicative of a "thumb on the scales" than today's distribution is (see chapter 11). In short, one cannot infer defects in the *processes* leading to occupational distributions simply from the presence of statistically disproportionate *outcomes*.

The drive to increase the representation of women in underrepresented fields seems to assume an unlimited supply of female workers, rather than recognizing that changes are largely zero-sum. If more women become engineers, physicists, carpenters, and bricklayers, fewer women will become biologists, doctors, nurses, and psychologists. Implicit in the argument for "underrepresentation" is the notion that some women have made the "wrong" choice or, at least, had the wrong choice imposed upon them.

The implications of this chapter are likely to be misunderstood. Does the fact that males and females differ, on average, in occupational interests and abilities mean that vocational counselors should use this information to steer females into certain occupations and males into others? Should employers routinely assume that all male and female applicants share the average characteristics of their sex? Certainly not in either case. There is sufficient overlap that it would generally be inappropriate, even if statistically rational in the absence of other information, to make *individual* decisions based on these generalizations. The existence of these differences does mean, however, that reflexive suspicion of statistical disparities is misplaced. That suspicion is warranted only if statistical parity is a reasonable starting assumption, which it manifestly is not.

6 The Gender Gap in Compensation

The average full-time female employee has approximately three-quarters the earnings of a full-time male employee. Few students of this "gender gap" believe that most of it is due to employers' paying women less than men for performing the same work.

Most of the contributors to the wage gap are unsurprising. Men tend to have more job-related education, training, and experience, and they tend to work substantially more hours than women, even within occupational categories. Greater mathematical ability—disproportionately possessed by men—also adds value in today's workplace.

Men and women tend to value different attributes of jobs, with men attaching more importance to financial success and women attaching more importance to helping others. Women place more weight on having flexible hours, shorter commutes, pleasant physical surroundings, and safe working conditions. Because an employer must generally pay employees more to do risky and unpleasant work than safe and pleasant work, these differences in preferences have economic consequences. Women also appear less likely to bargain hard for higher salary and are less likely to seek out and possess information relevant to the bargaining process.

Like the glass ceiling, much of the wage gap is an effect of marriage and families. Single women without children have earnings roughly comparable to single men, but married women with children earn only about 60 percent of married men's pay. Thus, the primary gap is not between men and women but rather between women with children and everyone else. Mothers, on average, work fewer hours and have more and longer extended absences from the labor force. Unlike women, men tend to work longer hours after becoming parents.

The wage gap will continue to close if women's work-force behavior increasingly resembles men's, although there is reason to doubt that it will ever disappear. It is questionable, however, whether equalization of earnings of men and women is a sensible policy goal, as long as women tend to trade higher pay for shorter hours, lower risk, and more pleasant working conditions.

The term *gender gap in compensation* is a shorthand phrase describing the fact that full-time female employees on average earn less than full-time male employees. The most commonly reported value for the "gender gap" is the female-to-male annual earnings ratio, which in 1999 was approximately 0.72, indicating that the average full-time female worker earned seventy-two cents for every dollar earned by a male. If weekly or hourly earnings are measured, however, the disparities are smaller, reflecting the fact that women work slightly fewer weeks per year and fewer hours per week. Thus, in 1999, the weekly earnings ratio was 0.76, and the hourly ratio was almost 0.84.[1] Demonstrating the progress that women have made in the workplace, the annual-earnings figure in 1977 stood at just fifty-nine cents. The gap narrowed at the rate of approximately 1 percent per year from the mid-seventies to the early nineties.[2] The annual ratio, but not the weekly or hourly ratio, has declined somewhat from its high of over 0.74 in 1997, in large part because of the entry of unskilled women into the workplace as a consequence of welfare reform.

Before turning to the factors that legitimately contribute to compensation disparities, we will briefly examine the widely reported but empirically unsupportable belief that the gender gap is due largely to discrimination.

Does the Gender Gap Indicate Discrimination?

The gender gap is often held up as obvious proof of wage discrimination against women. It is invoked to support toughening of the laws against sex discrimination and to demonstrate the need for sex-based affirmative action. President Clinton declared that complete parity is required: "75 cents on the dollar is still only three-quarters of the way there, and Americans can't be satisfied until we're all the way there."[3] Therefore, he asserted, the Equal Pay Act must be strengthened. Most knowledgeable people, however, acknowledge that the bulk of the gender gap is not caused by discrimination, although the size of the portion that might be due to discrimination is hotly contested.[4]

Perhaps the best evidence that the gender gap is not caused in any substantial part by wage discrimination in the ordinary sense is the fact that most of the pay gap occurs *across* occupations rather than within them. Within occupations, most of the remaining difference occurs *across* firms.[5] Thus, the overwhelming majority of the pay gap is due to women's being disproportionately employed in lower-paying occupations and by lower-paying firms.

One common method of testing for discrimination is *multiple regression analysis,* which allows analysis of the question whether, all else being equal, men still earn more than women. The first step is to identify factors (independent variables) believed to influence earnings (the dependent variable).[6] These factors differ across employers and across jobs. A comprehensive model would include measures of "human capital," which is defined by economists as investments in oneself, such as "education, health, on-the-job training, job search, geographic mobility, and other activities that enhance market earnings."[7] The model would also include various measures of productivity and working conditions.

A very simple regression for a particular job might include just seniority, years of education, and job-performance ratings. If these are the only variables that influence earnings, then knowing an employee's seniority, education, and performance ratings would allow prediction of his or her compensation with only random errors. If other important factors contributing to compensation were not included in the regression analysis—for example, number of hours worked per year—the regression would predict inaccurately. The more truly independent variables that are included, the more accurately one can predict earnings.

In testing for sex discrimination, a *dummy variable* for "sex" is included.[8] All of the variance in wages that is not accounted for by the other independent variables is assigned to the dummy variable. If the coefficient of the dummy variable is statistically significant, that indicates that sex and wages are somehow related, although it does not reveal exactly how. It could mean that compensation is directly influenced by sex (in the form of discrimination), or it could mean that wages are influenced by some omitted variable that is itself correlated with sex. If, in the above example, the number of hours worked per year has an influence on earnings independent of job-performance ratings—and if men work more hours on average than women—then what appears to be an influence of sex would in fact be an influence of the omitted "hours worked" variable.

Because some variables are difficult or impossible to quantify, regressions often incorporate *proxies*, that is, alternative measures that are believed to approximate the variable of real interest. For example, if "years of experience" is thought to be an important predictor of compensation but direct information about it is difficult to obtain, "age" or "years since completing degree" might be used as a proxy on the ground that they will be correlated with experience. In testing for sex discrimination, however, use of either proxy would skew the analysis in favor of a finding of discrimination because women are more likely than men to have interruptions in their work lives.[9]

Although numerous studies have been conducted that explain part—indeed most—of the gap, some disparity remains unexplained in virtually every study. That residual discrepancy may reflect discrimination or it may reflect the influence of unmeasured, or even unmeasurable, characteristics; the regression model cannot distinguish between these causes.[10] One must have an extraordinary level of faith in the wage models, however, to conclude that the residual is usually a measure of discrimination rather than inadequate specification of relevant variables.

A study of lawyer compensation is revealing. It controlled for graduating class, practice type (large firm, small firm, corporation, public agency, etc.), specialty (criminal, tax, etc.), and family status.[11] Notwithstanding control for these variables, the study revealed a substantial unexplained compensation gap between men and women, which the author attributed to discrimination. Astonishingly, however, the study did not control for two of the most significant contributors to lawyer income—hours worked and, in private firms, value of business brought into the firm. Given the well-known distaste that many female lawyers have for the long hours required in many law practices,[12] only the most credulous of observers could conclude on the basis of that study that female lawyers are victims of wage discrimination.

A recent National Science Foundation study of the wage gap among engineers showed how important one simple variable can be.[13] Not surprisingly, the gap among engineers was smaller than the overall gender gap, since the focus on engineers controlled for occupation, at least at a gross level. Female engineers earn approximately 87 percent what male engineers earn. The study found a number of differences between male and female engineers: females were less likely to have engineering degrees, slightly more likely to work for government and less likely to work for private industry, more likely to be computer software engineers and less likely to be electrical or electronics engineers or mechanical engineers, and less likely to have earned a Ph.D. Female engineers had, on average, five years' less experience than men, using "years since bachelor's degree" as a proxy for experience. By far the variable contributing most to the observed disparity was experience. When just that variable was included, the wage gap shrank from thirteen cents to three cents. Inclusion of the remaining variables shrank the gap by just an additional penny.

Should we conclude from the NSF study that employers cheat women out of that additional two cents? Not unless we are confident that the variables used represent the universe of legitimate factors affecting compensation. The report cautioned that the remaining difference was small compared to sources of potential errors in the wage model. Use of "years since bachelor's degree" as a variable would tend to overstate women's experience, and the study did not include a number of variables that might be relevant: degree combinations, worker productivity, quality of school from which degrees were received, quality of the employer, or "lifestyle and family-related choices," presumably including number of hours worked. Those variables could easily make up the remaining two cents or even more.

An earlier NSF study examining earnings by women in science and engineering had also failed to control for many potentially explanatory variables, but that report was less cautious in its conclusions.[14] The study reported that men holding doctoral degrees in science and engineering earned on average $13,300 more than women. Because the grouping of all scientists and engineers is more heterogeneous than the category of engineers (especially since "scientists" included social scientists), the gap was predictably larger than in the engineering study: women earned seventy-nine cents to the men's dollar. The large number of variables included in the wage study reduced the gap by 90 percent. The remaining 10 percent, according to the study, "can be interpreted as an estimate of employer preferences for different types of employees,"[15] in other words, sex discrimination. The only caution the report provided was that the figure is "at best, a rough estimate because statistical methods are never able to capture with complete accuracy the true complexity of human behavior." The message being sent was clear, however: female scientists and engineers are victims of sex discrimination. That is also the message that was received.[16]

The report's willingness to draw even tentative conclusions about employment discrimination from the NSF data is striking given other information in the report. On page 70, the report indicated that women science and engineering faculty engage in less research and publish less than men. On page 73, it stated that number of publications is a frequently used measure in analyzing academic salaries and that it would be "interesting" in the future to control for publications. On page 75, the report concluded that the differential in earnings in a study that did not control for productivity, despite acknowledged sex differences, is a rough estimate of employer discrimination.

The report also created a misleading impression about the size of the unexplained wage differential. By reporting the percentage of the gap that remained unexplained rather than the unexplained absolute wage difference, the report leaves the reader with the impression that the unexplained portion is larger than it is. Ten percent may sound like a substantial amount, but the remaining gap was not 10 percent of total wages but 10 percent of the initial twenty-one-cent difference. That is, what was left unexplained by the NSF study was the fact that, controlling for some variables but omitting productivity—which as we will see later in this chapter varies substantially by sex—female scientists and engineers earn approximately ninety-eight cents to the man's dollar.

No one can prove that no part of the gender gap is caused by discrimination, just as no one can prove that all, most, or perhaps even any of it is. However, a large portion of the gap is a result of forces that most people would not view as morally problematic. Indeed, given the large number of average differences between men and women in workplace behavior, it would be staggeringly surprising if men and women had equal earnings.

The Varied Contributors to the Gender Gap

If discrimination is not the primary contributor to the gender gap, what is? The answer is that there are a great many factors, many having only relatively modest effect by themselves but in the aggregate adding up to the one-third greater wages earned by men. Many of the contributing factors are relatively straightforward, and, like contributors to the glass ceiling, many appear to reflect psychological sex differences. In general, men tend to invest more of themselves in the workplace in order to attain both status and resources; women tend to invest more of themselves in their families and less in the workplace. Men earn more in large part because they tend to work more hours, occupy riskier jobs, work in less-pleasant environments, obtain greater job-related education and training, have fewer absences from work and fewer extended withdrawals from the work force, and possess a larger measure of the mathematical and spatial ability that is heavily rewarded in the modern age.

Although the gender gap in compensation is defined solely by wage comparisons, not all compensation is monetary. A substantial portion of compensation for many workers comes in the form of fringe benefits such as pensions and health insurance. Much of the compensation gap may disappear if fringe benefits are included in the earnings analysis. In a study based upon data from the 1991 National Longitudinal Survey of Youth, researchers found that while the average wage rate of full-time female employees between the ages of twenty-six and thirty-four was 87.4 percent of the average male rate, when fringe benefits were included the figure rose to 96.4 percent.[17] These figures do not take into account the fact that fringe benefits are often more valuable to women than to men because men and women make equal pension contributions but women tend to collect their pensions for more years, and because women consume substantially more health-care services than men.

Hours

Virtually all studies of labor-market behavior find large sex differences in hours worked even among full-time workers. That is why the gap in hourly earnings is substantially smaller than the gap in annual earnings. Most studies indicate that full-time male employees work approximately 8 to 10 percent more hours than females.[18] Full-time female employees are substantially more likely than male employees to work less than forty hours per week.[19]

At the high range of hours, the disparity can be even greater. Economist Victor Fuchs reported that among white married women with eighteen years or more of schooling and at least one child under twelve at home, less than 10 percent worked more than 2,250 hours per year, whereas half of their husbands did.[20] Moreover, one-third of the men worked more than 2,500 hours per year. A recent study of managers in an unidentified "large professional services firm" found that six times as many female managers than male managers had spouses who worked more hours than they did.[21]

Even in relatively homogeneous groups, the male tendency to work longer hours is apparent. For example, among MBA graduates from Stanford for the years 1973 through 1985, men averaged approximately 18 percent more hours per week than women (55.7 hours versus 47.1).[22] The influence of family is apparent, because single women did not differ from single men in hours or earnings. For holders of MBAs, as for most other workers, working more hours tends to result in higher earnings.[23]

Because of the sex difference in hours worked, the hourly earnings gap (16 cents in 1999) is a better indicator of the sexual disparity in earnings than the annual figure. Even the hourly earnings ratio does not completely capture the effects of sex differences in hours, however, because employees who work more hours also tend to earn more per hour. Therefore, a proper adjustment for hours worked would reduce the gap even further.

Often not captured in inquiries about sex differences in hours is the sex difference in absenteeism, which is substantial.[24] Although absences should not directly affect earnings if covered by employers' leave policies, frequent absences are likely to have indirect effects, such as interference with accumulation of work experience, reduction in employer investment in on-the-job training, and reduction in the prospects for wage increases and promotion.

Risk

Jobs differ substantially in the amount of associated risk, and, as we have seen before, attitudes toward risk are not distributed randomly with respect to sex. A look at the work force immediately reveals that men are engaged in the riskiest jobs; indeed, a list of the most dangerous occupations is a list of overwhelmingly male-dominated jobs—fisherman, logger, airplane pilot, metalworker, coal miner, and so forth. In any year, over 90 percent of workplace deaths are men. The higher the proportion of female employees in an occupation, the less likely it is that the occupation involves hazardous (or other onerous) working conditions.[25] Although women are increasingly represented in some traditionally male risky jobs, most of these jobs remain overwhelmingly male.

The relationship between attitudes toward risk and compensation is not limited to physical risk, for as we saw in chapter 4, men are more likely than women to take actions entailing "career risk." Career risks, like physical risks, operate in accordance with the general rule that with greater risk comes the possibility of greater reward. A study of female executive compensation has revealed that ambition and willingness to take risks are positively related to compensation.[26] Linda Subich and her colleagues have found that men's greater risk taking is related to greater salary expectations and may therefore be a factor in determining the probability of obtaining a higher salary.[27] Moreover, men are more likely to be employed under wage schemes having a greater component of pay that is contingent on performance, such as sales commissions and performance bonuses, meaning that they bear more of the risk of short-run variations in their performance.[28] High risk takers are more likely than the risk-averse to exhibit interest in commission sales jobs.[29]

The higher wages associated with risky jobs reflect what economists call "compensating differentials." The proposition dates back at least as far as Adam Smith, who wrote:

> The whole of the advantages and disadvantages of the different employments of labor and stock must, in the same neighborhood, be either perfectly equal or continually tending to equality. If in the same neighborhood, there was any employment evidently either more or less advantageous than

the rest, so many people would crowd into it in the one case, and so many would desert it in the other, that its advantages would soon return to the level of other employments.[30]

Wages are not simply a return on human capital investment—for special skills or training, for example. They are also compensation for unpleasant aspects of a job.[31] If two jobs are identical in their human-capital requirements but one carries with it a substantial risk of physical injury or uncomfortable working conditions, the risky or unpleasant job will usually carry a higher rate of pay. As Smith recognized, the more "advantageous" job (the safer one) will have more takers, thereby increasing the supply of those eager to take that job and decreasing the supply of those willing to take the more unsafe job, driving the wage of the safer job down and the wage of the more dangerous job up.

Many of the low-paid jobs occupied by women are low paid in part because they have desirable characteristics such as safety, flexible hours, and higher fulfillment, and are therefore in higher demand. Warren Farrell has pointed out that occupations in which more than 90 percent of the occupants are women almost always have at least seven of the following eight characteristics: ability to "check out" psychologically at the end of day, physical safety, indoor, low risk, desirable or flexible hours, no demands to relocate, high fulfillment relative to training, and contact with people.[32] Economist Randall Filer has concluded that a substantial portion of the wage gap can be explained by the fact that men tend to take jobs that are less attractive in some way than those filled by women.[33] That is, women give up some amount of wages in exchange for other attractive job attributes, so that, in Adam Smith's terminology, the "whole of the advantages and disadvantages" are equal.

Some investigators have questioned the extent to which compensating differentials actually can be demonstrated to operate in the labor market. The evidence in favor of compensating differentials is clearest for dangerous jobs, but there is substantial evidence in other contexts as well.[34] One difficulty is in quantifying unpleasant working conditions sufficiently accurately to include them in a wage model, and, indeed, given the heterogeneity of tastes, knowing whether particular attributes are positive or negative to workers.

Job-Related Education, Training, and Experience

Women as a group have a lower level of job-related schooling and work experience, although the sex difference is shrinking, and when occupation-specific training is controlled, much of the gap disappears.[35] Differences in college major account for a substantial portion of the wage gap, with the payoffs to degrees in engineering and business being greater than payoffs to degrees in education.[36] Part of the difference in human capital is due to the fact that in older cohorts of female workers many never expected to spend an extended period of time in the work force. For example, only 28 percent of white women ages fourteen through twenty-four in 1968 reported that they planned to be working at age thirty-five. When these women actually reached thirty-five, more than 70 percent of them were working.[37] Women who expected to be in the work force at thirty-five earned approximately 30 percent more than women who had not planned to be. The disparity between expectation and actuality led to large numbers of undertrained women in the work force.[38]

The compensation gap has been shrinking over recent decades because women are making more accurate judgments about their future labor-force participation and because they are increasingly represented in high-skill fields.[39] Nonetheless, even the Glass Ceiling Commission, which was singularly unwilling to assign any causal responsibility for economic disparities to women's workplace behavior, acknowledged that "more men than women continue to earn the degrees and credentials that are now generally considered to be prerequisites for senior management positions in the private sector."[40] The commission noted that men are far more likely to receive master's degrees in business, while women's advanced degrees are heavily concentrated in education. Labor Secretary Robert Reich was surely being disingenuous when he complained in the second Glass Ceiling Report that "over half of all Master's degrees are now awarded to women, yet 95 percent of senior-level managers of the top Fortune 1000 industrial and 500 service companies are men."[41] This is a meaningless comparison, since there are few senior managers of either sex with master's degrees in education.

Does Mathematical Ability Add Up to Higher Salary?

In many wage studies, "years of education" are viewed as fully fungible. The implicit assumption is that the monetary return to a year's education should be the same irrespective of educational field. That is not the case, however, for the simple reason that some fields of study require more of some scarce but valuable attributes than others. In today's technological society, mathematical ability is such an attribute.

Economists Morton Paglin and Anthony Rufolo performed an extensive study of the 180,000 takers of the Graduate Record Exam in 1982 that demonstrates the importance of educational field.[42] The GRE, which consists of verbal (GRE-V) and quantitative (GRE-Q) sections, is taken by most applicants to graduate (but not business or professional) schools. Paglin and Rufolo found substantial differences in student performance on the GRE depending upon major. For example, physics majors had an average GRE-Q score of 680 (on a 200 to 800 scale) and GRE-V score of 514, whereas education majors earned scores of 442 and 440, respectively. Students seemed to select themselves into disciplines fairly accurately. In the highest GRE-Q interval, for example, 71 percent had majored in math, physical science, or engineering, while only 2.6 percent had majored in education or social science. In comparison, 37.4 percent of students in the lowest interval (scores of 300 to 400) majored in education or social science, while only 1.2 percent majored in math, physical science, or engineering.

The most interesting aspect of Paglin and Rufolo's study for our purposes is the light it sheds on earnings. They found a high correlation between the quantitative demands of employment fields and mean starting salaries of college graduates in the field. High-paying fields, such as engineering, the physical sciences, and computer science, draw the bulk of their employees from the high end of the quantitative-ability distribution, while low-paying fields, such as education, social science, and the humanities, draw a large proportion of their candidates from the low end. Paglin and Rufolo found that every 100-point increase in average GRE-Q score for a particular major yielded approximately $400 per month more in income (in 1982 dollars). The relationship between quantitative demands and earnings was so strong that quantitative ability alone accounted for 82 percent of the variance in earnings among different fields.

A similar analysis using the GRE-V showed no significant correlation between average test scores of a field and income. Paglin and Rufolo interpreted this finding to mean that verbal ability is not sufficiently scarce to yield a premium to those who major in a subject requiring higher than average levels of it. In some sense, however, verbal ability is as scarce as quantitative ability; that is, by definition only 10 percent of the population is in the top 10 percent of verbal ability or of quantitative ability. But there are probably many more jobs for college graduates that actually require someone at the highest levels of quantitative ability than there are that require someone at the highest levels of verbal ability. While people with high verbal ability may be able to do some jobs incrementally better than people with average verbal ability, there are many jobs that can be done only by people with high quantitative ability. Moreover, quantitative ability may not only be necessary to some jobs, it may be sufficient. Employers looking for employees high in quantitative skills may care little about personality and social skills, but employers looking for high verbal abilities may require that their employees possess a number of other traits as well.

Paglin and Rufolo's findings have obvious relevance to the gender gap in compensation in light of the substantial difference between male and female representation at the highest levels of quantitative ability. Calculating the effects of the sex difference in GRE-Q scores, they found that their equation, derived without regard to sex, predicted a monthly salary of $1,667 for males and $1,344 for females, for a female-to-male salary ratio of 0.81. That is, without taking into account the many other factors that influence the size of the gender gap, nineteen cents of the gap in starting salaries of college graduates was explained by the quantitative demands of the field. Within fields, starting salaries for males and females were very close, with a salary ratio of 0.97.[43] Again, this finding demonstrates that a major reason for sex differences in compensation is that men and women do different work.

These results shed some light on sex differences in academic salaries. Academic disciplines in which women are numerous—such as English, foreign languages, and psychology—tend to pay less than fields in which women are scarce—such as engineering, computer science, and economics.[44] Sociologist

Marcia Bellas has argued that this pay differential is, in large part, a form of discrimination. Fields with a high proportion of women tend to pay less, even controlling for nonacademic demand in the discipline, she argues, specifically because of the concentration of women.[45] However, a degree in drama (the lowest-paid academic discipline she examined) may not reflect the same inputs as a degree in dentistry (the highest paid).

Paglin and Rufolo's results should not be taken to mean that quantitative ability is necessarily the route to high earnings. Their data described earnings of new college graduates. At higher educational levels, verbal ability, in conjunction with certain temperamental traits, may be a surer path to high earnings. A high-verbal individual who goes to law school may expect higher earnings than a high-math individual who goes to graduate school in physics. Even in the corporate world, high technical ability will take one only so far. To go beyond a position that simply allows one a comfortable living, it is generally necessary to move to the business side, in which case both verbal ability and all of the temperamental qualities of the businessperson discussed in chapter 4 will be required.

Strength and Nurturance?

Paglin and Rufolo point out that the form of analysis they employed would also work for such attributes as physical strength. As with quantitative ability, individuals (and the sexes) vary in physical strength, and some jobs require substantial amounts of it and must compensate for it. Women's lesser upper-body strength limits their entry into many physically demanding jobs.[46] As between two jobs requiring no special skills, one requiring substantial physical strength should pay more than one that does not, since the supply of people able to do the job is substantially smaller. Thus, even without taking account of differences in working conditions, the unskilled job of garbage collector should pay more than the unskilled job of parking-lot attendant. Ninety-eight-pound weaklings of either sex simply cannot lift garbage cans weighing perhaps thirty to fifty pounds all day, every day; they can, however, just like the physically strong, sit in a booth and collect parking tolls. All else being equal, therefore, weaker men should have lower average earnings than stronger men because they would have a more restricted set of opportunities.

Perhaps if the analysis works for strength, which favors men, it should work for nurturance, which would favor women. A number of occupations—nursing, teaching in the early grades, day care, and social work, for example—involve nurturance. Are there similarly positive returns to nurturance? Barbara Kilbourne and Paula England have argued that occupations that require high levels of nurturing skills, which have a high proportion of women, actually suffer a wage *penalty*. That is, the demand for these skills in an occupation lowers rather than raises the associated wages. According to Kilbourne and England, employers channel women into jobs requiring interpersonal skills "to capitalize on the way sex role socialization hones these skills in women."[47]

There are a number of legitimate reasons why nurturance would not carry a wage premium and might even carry a wage penalty. First, if it is true, as Kilbourne and England assert, that nurturance has been inculcated into women generally, then it is not the kind of scarce quality that will command a premium, since by definition it will be quite common. That is, it fails to carry a wage premium not because it is associated with women but because it is associated with so many women (as well as some men).

Jobs in which nurturance is a substantial asset are also likely to be pleasant in a number of ways: emotionally satisfying (often precisely because of opportunities for nurturing), regular hours, summers off (in the case of teachers), pleasant working conditions, and physical safety. These attributes will attract sufficient candidates that the financial return need not be as high as it otherwise would be. Thus, it may not in fact be negative returns to nurturance, but compensating differentials for pleasant working conditions, that explain the lower pay in these positions.

Finally, and perhaps most fundamentally, a high level of nurturing skills may be useful in many positions, but the necessary level of nurturance may not be that high. While one must be relatively high in mathematical and spatial ability to be a mechanical engineer and relatively high in physical strength to be a garbage collector, one need not be relatively high in nurturance to perform the functions of a nurse or a teacher. One who does not have a high enough level of quantitative ability simply cannot be an engineer, and without a high enough level of physical strength one cannot be a garbage collector; other positive intellectual or personality traits will not make up for the lack of these abilities. In contrast, people with relatively low levels of nurturance can still be nurses or teachers, however, even good ones. What they lack in nurturance, they may make up for in, say, intellectual ability.

Productivity

Sex differences in productivity also contribute to the wage gap. Although productivity is often difficult to measure directly, in a series of studies in Israel and the United States, economists Judith Hellerstein and David Neumark found female employees to be less productive than male employees. In the Israeli study, they compared data from manufacturing plants to assess the productivity of workers and their relative wages. The relationship that they found could hardly have been closer. The ratio of female-to-male productivity was 0.65, indicating that women were 35 percent less productive than men, and the earnings ratio was 0.66. They concluded that the wage gap in these plants reflects a productivity gap rather than discrimination. In the U.S. study, Hellerstein and Neumark, along with Kenneth Troske, similarly found a lower level of productivity of women workers, but unlike in the Israeli sample, lesser female productivity explained only about half of the entire wage difference. They estimated that while women were 15 percent less productive than men, they earned 32 percent less. The difference between productivity and earnings was significant for "technical, sales, etc." workers, but not for "managerial/professional," and for predominantly male plants but not for predominantly female plants. This difference indicates that some factor, whether wage discrimination or some other unmeasured factor (such as women's employment in lower-paying occupations at a level finer than the occupational breakdown that they used) was operating as well.[48]

There are at least two exceptions—lying at opposite poles of the hierarchy of occupational prestige—to the general lack of direct information on worker productivity. These exceptions are piecework employees and academics. In both categories, it turns out, males are more productive than females.

Men performing piecework are regularly found to produce (and therefore earn) more than women. Studies in the shoe, furniture, and cotton-weaving industries found that men earned 10 to 13 percent more than women.[49] Political scientist Steven Rhoads quotes the manager of a British furniture manufacturing concern as saying that there was "no limit, I mean *no* limit" to the exhaustion that his male employees were willing to endure to make more money under the company's piecework system. Men, it seems, are more willing than women to push themselves to the limits of their endurance to increase their pay. The piecework differential is not a major contributor to the wage gap, of course, because piecework compensation is no longer widespread. It does, however, illustrate the point that men may tend to be more motivated to engage in intense labor-market activity than women.

Men performing academic work exhibit the same tendency toward greater productivity. Scores of studies of academic productivity over the past decades have consistently found that men write more than women do, typically about 50 percent more.[50] Sex differences in publication rates tend to be smaller early in the career and grow as scientists get older. The greatest portion of the difference is a result of especially large sex differences among the most productive. For both men and women, between 10 and 15 percent of scientists publish about half of all articles produced.[51]

More recent studies have continued to find reliable sex differences in academic productivity. A large-scale study of medical-school faculty found that women had published only about two-thirds the number of articles that men had in their careers (24.2 for women and 37.8 for men),[52] even though men and women differed little in experience. There was no effect of marriage on productivity in either sex, and at best a slight positive effect on men for having children. Nonetheless, women with children had less than two-thirds as many publications as men with children (18.3 versus 29.3).[53] Although sex differences in productivity are sometimes attributed to lesser "mentoring" of female academics, no sex difference in either the quality or prevalence of "mentoring" relationships was found in this sample. Interestingly, 80 percent of women faculty who had mentors said that it was not important to have a mentor of the same sex.[54] A nationwide study of academic departments of pediatrics likewise found that men not only worked more hours but had greater academic productivity in terms of publications and grants; women spent more time in teaching and patient care.[55] Studies of academic psychiatrists yielded similar findings.[56]

Article number is, of course, most directly a measure of quantity, not quality. Assessments of quality of research are much more difficult. One measure that is sometimes used is citation count—a measure of how often an author's work is cited in the literature. While citation count is not a direct measure of quality, it is a measure of visibility and perhaps of impact. Findings of sex differences in citation count are less consistent than differences in article number. Some studies find that articles written by men are more likely to be cited, some find that articles by women are more likely to be cited, and some find no sex difference at all.[57]

Some have argued that women may publish less than men because of discrimination in provision of resources to female scientists.[58] Examination of that question is difficult because of its chicken-or-egg nature. Funding, research space, and other support are simultaneously rewards for prior research and resources for future research. Thus, a simple correlation between funding and research productivity could reflect the fact that productive scholars are rewarded with these resources or that some people, perhaps disproportionately men, are provided research support that enables them to achieve higher levels of productivity, or both.

If differential productivity is a consequence of differential research support, then the productivity differential should diminish dramatically in disciplines where research does not depend heavily on institutional support. Such is the case with publishing by law professors. Most legal research is done in a library (either brick-and-mortar or virtual), and most of it is dependent upon neither grants nor laboratory space and equipment. Although data on sex differences in legal publishing are scant, there is some indication that the general tendency of men to out-publish women prevails in law as well. James Lindgren and Daniel Seltzer conducted a study of the most prolific law professors in the United States as measured by publication in the top twenty law reviews.[59] While not an ideal measure of either, the study is at least a crude measure of both quantity (number of articles within a limited sample of publications) and quality (articles appearing in the top journals). Among the top twenty-five producers, only one was a woman. Because the study was not designed to test for sex differences or to measure overall productivity, these findings provide only weak evidence for a sex difference in productivity among legal academics, however. Those interested in empirical study of a possible causal link between institutional support and faculty output might productively examine disciplines, like law, in which institutional support is less important.

What accounts for the sex difference in academic productivity? Scholarly productivity is the currency of academic status, even though other duties—such as teaching and service—are also required of most academics. Stephen Cole and Robert Fiorentine concluded on the basis of their literature review that male scientists outproduce female scientists "because it is more important to men to be occupationally successful than it is to women."[60] This statement does not mean that women care less about producing competent work than men do, but rather that it is more important to men to be recognized as being at the top of the heap, and as a result they are likely to devote a greater portion of their energies striving for higher levels of attainment.

Defining productivity of academics primarily in terms of scholarship admittedly biases productivity judgments against those individuals (disproportionately women) who devote the bulk of their energies to teaching and service to the institution rather than toward research. Given that everyone understands that scholarly productivity is the sine qua non of academic status, however, it is not obvious that this bias is an unfair one. From the perspective of research universities, the primacy of research productivity is virtually axiomatic. It is difficult to envision a university's valuing undergraduate teaching equally with research achievement, no matter what lip service it might pay to teaching. A brilliant scholar who is a competent teacher and a competent scholar who is a brilliant teacher are simply not of equal value to a university. The impact of the world-class scholar may be felt around the world, adding to the university's international luster, while the impact of a world-class teacher is felt primarily, though not exclusively, in the particular classroom in which he teaches.

Helping Others versus Helping Yourself

In addition to differences in what men and women put into a job, men and women also differ in what they are looking to get out of a job; thus, the indicia of a "good job" are often different for men and for women. Men are more likely than women to attach primary importance to financial aspects of the job and are thus more likely to choose careers on the basis of how much money they can make.[61] For both men and women, the importance assigned to financial success is positively correlated with earnings.[62] Men also rate higher than women opportunities for advancement, acquisition of influence, and freedom on the job.[63] Men are also more likely to view the opportunity to be a leader as important in selecting a career.[64]

Sex differences in competitive orientation have a direct impact on the world of work. In a study examining sex differences in attitudes toward work, Terence Martin and Bruce Kirkcaldy found, for example, that men substantially outscored women on competitiveness ($d = 0.98$) and valuation of money ($d = 0.95$), while women outscored men on "work ethic" ($d = -0.49$), which is a measure of motivation to achieve based on reinforcement in the performance itself.[65] Another study found that the strongest predictor of the importance attached to money was the motive to outperform others.[66] Thus, it seems, though women care as much or more than men about doing well in many contexts, men care more than women about doing better than the next guy.

Although economic returns are important to women—after all, why else would they or anyone else work?—they are often not as important as for men. Studies regularly find, for example, that women rate the opportunity to help others or to help society as being more important than men do.[67] While one might not expect that a desire to help others would result in lower wages, economist Robert Frank found in a study of Cornell graduates that "salaries fall dramatically with increases in employer social responsibility."[68] At one level, that result is not surprising, since a business major who goes to work for a large corporation would be expected to earn more than a social-work major who takes a job in a state social-service agency. However, the relationship found by Frank held even after controlling for sector of employment, as well as sex, curriculum, and academic record.

Other Job Attributes

Women attach greater importance than men to other nonwage aspects of the work, such as relations with co-workers and supervisors, freedom to take time off, shorter commute time, opportunities to work part-time, pleasant physical surroundings, and safe working conditions.[69] The differential preferences of men and women are reflected even within occupations. Among physicians, for example, women are less likely to specialize, and when they do they tend to concentrate in low-prestige specialties.[70] Among nurses, while over 95 percent of all nurses are women, 42 percent of nurse anesthetists—whose average annual earnings approach six figures—are men.[71] Despite assertions that "women are pushed into" traditionally female medical specialties,[72] there is little evidence that women have less choice over their specialty than men do. Nonetheless, men are more likely to enter private practice, while women are more likely to take jobs as salaried employees in nonentrepreneurial settings.[73] Men spend more time each week reading the professional literature.[74] Even within specialties these differences hold. Among obstetrician/gynecologists, for example, women are more likely than men to desire work in salaried positions with regular hours, and men are more oriented toward higher incomes and private practice.[75] Female ob/gyns are considerably more likely to work for health maintenance organizations and to work significantly shorter hours than men.[76] Female dental students have been found to have a stronger preference for working as an employee than males, and male students have a stronger preference for solo practice.[77] Similarly, among veterinarians—an occupation that appears well on its way to being predominantly female—women work fewer hours than men, are less interested in practice ownership, and tend to set their prices somewhat lower.[78]

Comparable sex differences are revealed in studies of pharmacists. Women more than men are drawn to pharmacy by the opportunity to help people, and female pharmacists are much more likely to be employees. Men are more interested in the financial and business aspects of pharmacy and are more interested in being solo owners than women are.[79] A detailed study of pharmacists in Canada found that despite parity in numbers, female pharmacists earned less than males.[80] Many of the usual forces were at work: women worked fewer hours, had experienced more extended leaves of absence, were more likely to work part-time, were more likely to work in hospitals than retail settings, were less likely to own their own stores, and were less likely to be managers in either the hospital or retail sector. In entering the profession, men were more likely to have been drawn by the prospect of self-employment and women by the possibility of flexible work arrangements. According to the women, they did not seek ownership positions in the retail sector because such positions would be too demanding on their time, since business owners are often faced with fifty- and sixty-hour workweeks, much of it in the evening and on weekends.

Curiously, the fact that men and women had different incomes within practice types was viewed by the researchers as "clear evidence of wage discrimination."[81] This finding (and the condemnation) extended even to the lower income of female pharmacy owners. If female pharmacy owners are victims of wage discrimination, presumably they are discriminating against themselves. So eager were the researchers to find female victimization, they interpreted the finding that male pharmacists had experienced more difficulty in securing a bank loan as suggesting that "either female pharmacists have never aspired to retail store ownership (or franchise holdership) or have learnt that banks are not particularly forthcoming with loans for women, and have accordingly diverted their aspirations away from entrepreneurial endeavours." While either of these suggestions may be true, the researchers did not consider the more obvious possibility that women might simply have been more successful in securing loans. There is little doubt that had their survey revealed that women reported more trouble securing a bank loan than men, the researchers would not have interpreted women's difficulty as reflecting a male disadvantage.

Pharmacy appears on its way to becoming a largely feminized profession, just like such occupations as book editing, public relations, baking, and insurance adjusting and examining before it.[82] In 1990, 32 percent of pharmacists were women; by 2000, that figure had risen to 46 percent, and about two-thirds of the nation's pharmacy students were women.[83] The increasing market share of large chain drugstores and the corresponding decline in the small-business model of pharmacy may make pharmacy a less attractive profession for the men (and somewhat fewer women) who were drawn to the profession by the prospect of business ownership.[84] If so, pharmacy will probably be held up as another example of an occupation that decreased in status and pay with the increasing participation of women, but the causal attribution will not be quite right. It may be true that the increase in proportion of women was causal in a sense, because it increased the pool of pharmacists eager to work as employees. However, the decline in pay and status will not be primarily a result of the increasing proportion of women but of the increasing proportion of employees, as well as a decrease in the average number of hours worked.

You Get What You Ask For?

Another potential source of earnings disparity is a sex difference in willingness to negotiate for higher pay. Although in lower-level jobs the wage rate is often fixed, at higher levels, salary is often subject to negotiation. It is quite plausible that women do not bargain as hard over salary as men do. If so, then women performing roughly the same job for an employer would tend to be paid less for equivalent work. While that might look like wage discrimination by the employer, it would in fact be the facially neutral practice of paying people the lowest wage they are willing to accept.

The widely replicated finding that women place a lower priority on salary than men would cause one to suspect a priori that women either ask for less or bargain less intensely for a higher amount than the employer initially offers. Moreover, sex differences in status seeking may also have an impact on salary. Concern about salary level is often not really about money but about one's place in the hierarchy. It would not be surprising to find that people who are more concerned about money and status ask for more and bargain harder to get it. A recent meta-analysis of negotiation studies confirmed this suspicion, concluding that males consistently negotiated better outcomes than females, irrespective of sex of opponent.[85]

A well-known study performed by Charlene Callahan-Levy and Lawrence Messé suggests that a female tendency to "settle for less" appears relatively early in development.[86] They recruited subjects, ranging from first grade to college, to work for pay and then had them allocate pay to themselves and others. While both males and females tended to overpay themselves, males overpaid themselves by a greater amount and tended to view a greater amount as fair. The researchers concluded that one reason that females appear more generous—that is, by allocating more money to others—is "their indifference, relative to males, to receiving 'appropriate' money for their work." Even the first-grade girls tended to underpay themselves relative to boys. Interestingly, girls with more masculine occupational preferences tended to pay themselves more than girls with more traditionally feminine preferences. The researchers concluded that part of the gender gap in compensation could simply be the result of women's willingness to work for lower wages than men.

Suppose that the Callahan-Levy and Messé study accurately reflects female workplace behavior: when asked how much salary it would take to get them to accept a job offer, they systematically name a lower amount than a man would, or when offered a particular amount they do not negotiate as hard for a higher amount. Men and women with equal qualifications would then have systematically different earnings. Indeed, there is evidence that suggests just such an effect. Brenda Major and colleagues have found that women do tend to expect less in wages than men expect and, in a laboratory study, they found that the more money an applicant asked for the more the applicant received if hired.[87] There is also evidence to suggest that men are more likely than women to ask for a raise, rather than waiting for one to be awarded.[88]

Salary differences that are a consequence of differential bargaining would show up as part of the unexplained residual in a wage regression, but is it discrimination? The employer is not engaging in intentional discrimination—that is, the practice is facially neutral (unless, of course, the employer systematically engages in harder bargaining with women). Arguably, the practice of paying people what they ask for may be a species of "disparate impact" discrimination—a neutral practice having a differential effect on the two sexes—but it is not clear that compensation systems are subject to sex-based disparate-impact challenges. Even if they were, courts may not view an employer's practice of paying people whatever they ask for within some acceptable range as the kind of compensation practice subject to challenge. After all, the employer's defense in such a case is "I paid her what she asked for and now she is suing me because somebody else asked for more."

Information is critical to effective bargaining over salary, or anything else for that matter. There is evidence that women are less likely than men to possess relevant occupational information when engaging in a job search, and they are less likely to seek it out when they do not have it.[89] This behavior is, of course, consistent with a lesser centrality of wages, on average, to women.

If it turns out that differential bargaining is a contributor to the wage gap, it would seem that the only practical way to reduce its effect is to encourage women to value their services more highly and bargain more intensely over salary. It is difficult to see what an employer could do to eliminate this effect, short of setting a fixed salary for a job and refusing to bargain with anyone.

The Effect of Families

Much of the wage gap, like the glass ceiling, is related either directly or indirectly to marriage and families. Single women without children earn over 95 percent of single men's pay, while married mothers earn only 60 percent of married men's pay.[90] Among individuals between the ages of twenty-seven to thirty-three without children, women earn approximately 98 percent of what men earn.[91]

Although marriage is not consistently found to have an effect on women's earnings, motherhood generally has a negative effect.[92] Why does having children decrease women's earnings? Women with children work fewer hours than women without children and are more likely to be intermittent workers, leaving the work force for an extended period while their children are young. After an extended absence from the labor force, wages upon reentry for both males and females tend to be lower than upon departure, and the longer the period of withdrawal, the greater the decline in wages.[93] Solomon Polachek has found that absences from the labor market may explain as much as 93 percent of the gender gap.[94]

Unlike the case with women, married men—and especially married men with children—tend to earn more than single men.[95] This finding is simply correlational and does not by itself show that marriage actually causes men to become more productive. Some have argued that the causal arrow may point in the opposite direction—the traits that cause men to be productive may also cause them to be more attractive as husbands and therefore more likely to marry.[96] Such an effect would certainly not be surprising given what we know about female mate preferences. Economists Sanders Korenman and David Neumark have expressed skepticism about that interpretation, however, in light of the lack of relationship between wage growth of unmarried men and the probability that they marry.[97] If the earnings advantage that these men enjoy after marriage were simply a reflection of their inherent qualities, rather than their situation, one would expect to find reflections of it in their premarital wage profiles. Korenman and Neumark also found that wage growth occurs after, rather than before, marriage, although that finding is also consistent with women's recognition of the men's potential for high earnings.

Although the rate at which women leave the labor force has been decreasing in recent years, women still leave the labor force at a rate approximately three times that of men.[98] These hiatuses may have a significant effect on subsequent earnings.[99] Not all of these gaps are due to child rearing, however. Sociologist Karen McElrath found in a sample of academic criminologists, for example, that while women were four times as likely as men to leave the work force for at least a year, three-quarters of these changes were due to spouse's employment rather than maternity.[100]

Part of the wage gap is related directly to the allocation of domestic responsibilities in the home. Household responsibilities of women affect women's wages by decreasing both human capital investments and the amount of effort available for market work.[101] As a general rule, as time spent in domestic activities increases, earnings decrease.[102]

The effect of marriage and family on women is often felt even before they are married. Many women planning for their future careers take into account the potential future work-family conflict in making decisions about acquiring human capital.[103] Thus, women who expect or want to be stay-at-home mothers are less likely to invest in education and training that is directly relevant to the workplace. It seems likely, for example, that college women planning to get married and stay home with their children are more likely to major in, say, art history or French literature and less likely to major in business or science than women who either do not plan to have families or who expect to pursue active careers along with their families.

Career Priorities of Dual-Career Couples

Most couples place a higher priority on the husband's career than the wife's. One reason is that husbands tend to be more involved in their careers than their wives are.[104] Another is that husbands tend to earn more than wives, not only because men tend to earn more than women, but also because husbands tend to be older than their wives and therefore to be more advanced in their careers. Even if the wife earns more than the husband, however, priority still may go to the husband's career, either because of the likelihood that the woman will leave the work force in the future or because the couple believes that their collective emotional well-being is better served by the man's career success than the woman's.

The greater valuation of the husband's career imposes constraints on the wife's earnings. Economist Robert Frank found that priority of the husband's career can lead to wives' being overqualified for their jobs.[105] Because husbands tend to work more hours, possess larger stocks of human capital, and earn more, the location that maximizes the couple's income will tend to favor the husband's job, requiring the wife to make career compromises. The effect may be even greater than an income-maximization practice would imply, however, because even couples in which the husband does not earn more than the wife often give priority to the husband's career.[106] Because men's self-esteem tends to be more intimately connected with workplace success than women's, the couple as a whole may decide that it will profit psychologically by relocating for a job that is a notch higher for the husband and a notch lower for the wife, even if the couple's total income is unaffected.

Frank's suggestion that a couple is more likely to move for the husband's job than for the wife's finds ample corroboration in the literature.[107] A study of Ph.D. ecologists, for example, revealed that 35 percent of married women but only 11 percent of married men had moved to accommodate the relocation of their spouse.[108] A related fact is that in searching for jobs, women are more likely to face geographic constraints in their job search than men are. A study of law professors, for example, found that women were twice as likely to report that their search had been geographically constrained (approximately one-half of women and one-quarter of men reported such constraints).[109] In a field as specialized as law teaching (there are only approximately 175 law schools in the country, and most will probably not be hiring in any particular candidate's field of expertise in any year), geographic constraints not only may mean settling for a lesser job than one would otherwise obtain, but they may mean the difference between finding a job in the field and not finding one at all.

One must be careful in drawing causal conclusions from sex differences in behavior such as relocating. It is clearly the case that women are less likely to relocate than men and that couples are less willing to relocate for the wife's career than for the husband's. It is nonetheless possible that employers discriminate in offering relocation opportunities. Indeed, the Glass Ceiling Commission reported that women are not asked to relocate as often as men, a fact that the commission suggested may prejudice women's chances for advancement.[110] Unfortunately, as is true of almost all of the statistics reported by the commission, there was no control for the kind of position held. Given women's concentration in staff positions and in the government and nonprofit sectors, a sex disparity in relocation requests would be expected.

The Future of the Wage Gap

The gender gap in compensation will continue to close, as women's work-force behavior becomes increasingly similar to men's.[111] The average experience of employed women continues to grow, and women continue to invest more heavily than ever before in human capital, as evidenced by the dramatic increase in their representation in business and professional schools over recent decades. The decreasing rate of marriage may also have an effect on the gender gap in light of the negative effect of marriage on women's wages and the positive effect on men's. These trends may diminish the gap over the course of the next generation or so, but they will not eliminate it, except in the unlikely event that the changes are so dramatic as to make women indistinguishable from men in the traits that lead to high income.[112]

Complete convergence seems unlikely given the stability of sex differences in job preferences. A recent review of national surveys of adult workers from 1973 to 1994 found, across age groups and parental status, consistent sex differences in three job attributes: desire for a sense of accomplishment (favoring women) and a desire for promotion opportunities and for job security (favoring men). Contrary to the researchers' expectations, the data revealed that the differences actually appear to have widened in recent years, especially among younger workers.[113] Younger men especially are now attaching greater importance to high incomes and less to a sense of accomplishment than they did in the past. By mid-career (ages 36 to 50), these differences have shrunk considerably, demonstrating the importance of specifying the stage in the life course when making statements about men and women's preferences. For those who would like men and women to follow the same career trajectories, these data must be disappointing, because not only has there been no decline in sex differences in job-attribute preferences, the greatest differences are found within the cohort of workers just embarking on their careers.

It is occasionally asserted that the persistence of the gender gap is a measure of societal opposition to equality for women.[114] In fact, however, it is because of, rather than in spite of, the increase in women's labor-force participation that the gap did not close faster. If only young, highly skilled women had been entering the work force in the 1960s and 1970s, the gap would have narrowed more rapidly. Instead, large numbers of women who had never expected to be permanent participants in the labor force joined it and were paid in accordance with their skills and experience, a trend reprised on a smaller level when the welfare laws were changed in the mid-1990s to require recipients to work. This point can be made starkly by considering the fact that for most of the 1950s, the female-to-male annual earnings ratio was at least 0.63, and in 1960 it was almost 0.61. Yet after 1960, the ratio did not once again get as high as 0.60 until 1976, and it has only been since 1982 that it has consistently maintained itself above 0.60.[115] Thus, for almost all of the 1960s and 1970s—decades of extraordinary expansion of the role of women in the economy—the earnings ratio was *lower* than it was in the 1950s. For those who believe that the earnings ratio reflects the extent of opportunities for women, this is a difficult statistic to explain.

Some factors that contribute to the wage gap lie largely outside of the labor market. Approximately two million men in the United States are currently incarcerated, and the bulk of them are relatively low in job skills and workforce attachment. If these men were reintroduced into the labor force, the relative position of women in the economy would improve if only through the increased number of low-paid men.

The fact that men and women make different workplace choices might be surprising if they behaved identically outside the workplace, but sex differences in workplace behavior are no larger, and perhaps smaller, than sex differences in leisure behavior. Indeed, the strongest predictor of adolescent leisure activities is sex.[116] Sex differences in leisure activities, which are found cross-culturally,[117] are presumably beyond the influence of employers. They are, however, not unrelated to differences in workplace behavior. There is, for example, a high correlation between participation in highly competitive leisure activities and competitive work attitudes.[118]

It is difficult to disagree with economist Jennifer Roback's assessment that "the emotional intensity surrounding the discussion is . . . out of proportion to the magnitude of the problem as measured by the data."[119] The simple observation that men and women have different average earnings tells us very little about justice or fairness or equity once we recognize that people trade wages off against other desirable job attributes and life interests. Because earnings are easier to quantify and compare than other important job attributes, excessive attention has been paid to wage disparities. If we seek "equality as identity," however, we have to talk about "death gaps," "pleasantness gaps," and "hours gaps," as well as "wage gaps." Yes, women earn less, on average, but men die on the job more, work in less pleasant environments, and work more hours; if women have a cause for complaints about equality, presumably men do as well.

Once it is acknowledged that people make tradeoffs between money and risk and working conditions and time, it must also be acknowledged that it is folly to focus on just one dimension in making our judgments about equity. Instead, the entire constellation of job characteristics and rewards must be assessed. If some people select low wages, low risk, pleasant conditions, and short or flexible hours, it is hard to argue that they have a moral entitlement to the same wages as those occupying high-risk jobs with unpleasant working conditions and long hours. While we would all like to be able to choose the work, working conditions, and pay as if from a Chinese menu, the realities of economic life dictate that the entire bundle must be either accepted or rejected. In other words, one cannot choose the work and lifestyle of a schoolteacher and the salary of a brain surgeon.

III The Proximate and Ultimate Origins of Sex Differences

The sex differences in workplace outcomes described in part II follow in large part from the sex differences in temperament and cognition described in part I. Before embarking on a discussion of the policy implications of these differences, it is useful to consider just where these differences come from.

Despite fairly broad agreement on the existence of some sex differences, there is a decided lack of consensus about their causes. More sociologically minded commentators tend to view the sexes as largely equivalent and see sex differences as products of differential conditioning. Those with a more biological orientation tend to doubt the strong version of the sociological account—that the human brain is sexually monomorphic and that differences in temperament, cognition, and behavior are solely a consequence of different inputs into these sexually monomorphic brains. While not doubting the potential importance of social influences, they see these influences as acting on a sexually dimorphic brain, so that even with identical social conditioning, sex differences would be expected to remain.

The discussion of origins can easily get bogged down in sterile and futile debates over familiar, but false, dichotomies—learning/instinct, nature/nurture, cultural/biological—dichotomies that have substantially impeded understanding of the origins of human behavior. The ability of young children to soak up language is well recognized, but it makes no sense to discuss whether learning a language is cultural or biological. The predisposition to learn language is, of course, innate; that is why humans can learn languages and other animals cannot. For language acquisition actually to occur, however, the child must be exposed to language at a specific period in development, and it is the particular language to which the child is exposed that the child learns. Especially in an intelligent social animal such as humans, even biologically predisposed behaviors often require substantial social inputs to be realized.

The next chapter will discuss why the familiar argument of socialization, though often invoked, cannot fully explain existing sex differences. This is not to deny that there are differences in the social environments of boys and girls, but those differences are often evoked by children rather than unilaterally provided by adults. Moreover, there is too much that the socialization argument cannot explain, such as the early appearance and cultural universality of many of the differences.

The remaining two chapters present evidence for biological causes. Biologists distinguish between two types of causes: *proximate causes* are more immediate causes that are often explained in terms of biochemistry or physiology, while *ultimate causes* refer to the adaptive function of a trait or behavior.[1] The two kinds of causal explanations are not competitors but are complementary, as every trait has both proximate and ultimate causes. For example, if the question is why bears hibernate, proximate causes could be invoked—cold weather and a reduction in the availability of energy-rich foods—or an ultimate cause, or evolutionary reason, could be identified—the altered physiological state of the bear allows it to reduce its energy consumption in order to survive a harsh winter.[2] In chapter 8 I present evidence for the proximate cause of sex differences, which is primarily differences in exposure to sex hormones. In chapter 9 I discuss why, in evolutionary terms, male and female humans are, consistent with the general mammalian pattern, sexually dimorphic in both body and brain.

7 Why Socialization Is an Inadequate Explanation

It seemed clear to me that any between-sex differences in thinking abilities were due to socialization practices, artifacts and mistakes in the research, and bias and prejudice. After reviewing a pile of journal articles that stood several feet high and numerous books and book chapters that dwarfed the stack of journal articles, I changed my mind.

—Diane Halpern, *Sex Differences in Cognitive Abilities*

Sex differences, including the temperamental and cognitive differences credited with sex differences in workplace outcomes, are frequently, but implausibly, viewed as purely products of differential socialization by parents, schools, and peers. Temperamental differences appear before children can identify their own sex or the sex of others, however, and they are consistent cross-culturally. Male toddlers roam farther, have higher activity levels, and engage in more rough-and-tumble play. Sex-typed toy preferences precede children's knowledge of sex stereotypes, and same-sex playmate preferences often precede children's knowledge of what children are the same sex as themselves.

The extent of differential treatment of boys and girls by parents and teachers is often overstated, and the impact of children's temperament on parental treatment is often ignored. Complaints that teachers give boys more attention than girls obscure the fact that boys are more likely to volunteer answers than girls and that much of the extra attention received by boys is disciplinary in nature. Rather than being agents of a sexist socialization that results in boys acting like boys and girls acting like girls, teachers positively reinforce stereotypically feminine behavior in both sexes.

Among the three socialization agents, peers take the greatest role in policing "gender boundaries." The ease with which children seem to assimilate sex roles and their vigilance in policing them suggest that children may be "primed" to learn them just as they are primed to learn language. Rather than responding indiscriminately to feedback, children tend to be quite selective, being affected more by feedback for acting in a sex-typical way and more by feedback from same-sex peers.

Cognitive sex differences likewise appear early in life, although they are enhanced at the onset of puberty, and they are also found cross-culturally. Although levels of mathematical and scientific performance differ by country, the greatest sex differences are found in countries with the highest performance, suggesting that efforts to enhance math and science performance may actually increase sex differences.

The social factors generally blamed for the performance gap—such as a perception that math is for boys and pressure for girls (but not boys) to "play dumb"—

are inadequate explanations. Girls consistently get better grades in math throughout their schooling, a fact that is difficult to square with either of these arguments.

The socialization argument implies that in the absence of sex-typed societal expectations, the sexes would be largely identical in temperament, cognitive ability, and social behavior. The experience of the Israeli kibbutzim belies that argument, however. Despite an egalitarian ideology under which men and women were to do the same work and the creation of children's houses to relieve parents of care obligations, traditional sex roles quickly reemerged.

The familiar story that many of us grew up with is that the sex differences that we observe are purely social creations. Society, it is said, causes these differences by treating functionally identical individuals differently. Men are competitive, acquisitive risk takers, and women are cooperative, risk-averse nurturers because they learned these traits as little boys and girls and have been reinforced in them ever since. Boys are good at math because they are expected to be, and girls are bad at it for the same reason. These arguments have been difficult to refute in the way they are often framed, because of an underlying assumption that demonstration of social influences on sex differences is a complete refutation of biological influences.

Humans are an intensely social species, so it would be worse than foolhardy to assert that socialization is without influence. Sex is perhaps the most salient characteristic of individuals. We can be sure that sex is a highly salient characteristic for most other animals, as well. Even if we have trouble telling a male dove from a female dove, we can be confident that doves themselves have no such trouble. Even across species, sex differences can be highly salient. Dogs, for example, readily distinguish human males from human females, demonstrating greater defensive-aggressive reactions to men than to women.[3] The question for consideration here is not whether there are social influences on sex roles; instead, it is whether those influences act on a sexually monomorphic mind or on a sexually dimorphic one.

The central flaw of many of the socialization arguments is not that they invoke the environment as an important factor; no biological argument can do without environmental effects either. The flaw is the failure of most of the arguments to specify what part of the environment matters or why it matters. Invocation of some unspecified "socialization" is impossible to falsify, because ruling out one or more social forces does not rule out some other, unspecified, force. If the specific environmental factor thought to influence an outcome is specified, however, then the hypothesis is subject to testing and potentially to falsification. Even if the socialization hypothesis is supported, however, that does not mean that meaningful biological influences must be ruled out. The question then is what biological attribute of the organism causes it to respond in particular ways to specific stimuli. When viewed in that light, the question "is it biology or is it culture?" is revealed to be an empty one.

Now even the most ardent socialization advocates do not argue that biology is entirely irrelevant; after all, learning from the social environment requires a brain, a biological organ. But they do tend to assume that biology is not relevant in any interesting way. The SSSM, discussed in chapter 1, generally treats the brain as a general-purpose learning device, rather than as an organ that is predisposed to attend selectively to different portions of the environment and to respond in somewhat predictable ways. As we proceed through this chapter, the reader should ask why society is so effective at getting males and females to act in the way that it wants them to when it has so much trouble obtaining compliance with its other desires.

Temperamental Differences

The core of the socialization argument is that parents, teachers, and peers treat males and females differently, reinforcing sex stereotypes throughout their lives. Biologist Anne Fausto-Sterling argues, for example, that the difference in children's genitalia "leads adults to interact differently with different babies whom we conveniently color-code in pink or blue to make it unnecessary to go peering into their diapers for information about gender."[4] From the moment of birth, the argument goes, boys and girls are indoctrinated in society's expectations of sex-appropriate behavior. Usually assumed but not demonstrated is that these hypothesized differential expectations are not themselves products of biological predisposition on the part of those holding them and that such differential treatment as exists is actually the cause of differences in subsequent behavior.

The Early Appearance of Sex-Typed Activity

The early appearance of many sex differences makes their purely social origins suspect. Children generally cannot identify their own sex until two and a half to three years of age, and over the subsequent eighteen months to two years they learn to identify the sex of others.[5] Not until around age five do children begin to model their behavior after children of the same sex.[6] Yet many sex differences long precede these ages.[7]

Children are sensitive to sex differences long before they can assimilate society's expectations. Some infants apparently can discriminate between male and female adult faces as early as five months of age, and by twelve months some can distinguish between males and females even when hair and clothing are altered to make the sexes appear more similar.[8] They also seem to pay more attention to same-sex others in infancy.[9] Thus, the categorization of males and females is accomplished early by the infant mind.

One-year-olds show even greater sex differences than infants. Girls seem more reluctant to separate from their mothers and more eager to return when separated.[10] Even at this age, boys are more independent, more exploratory, and more active. Boys have a higher activity level than girls from the first year of life, and this difference persists throughout childhood.[11] Substantial sex differences in toy preference exist even in one-year-olds. In a study of one- to five-year-olds, effect sizes for sex-typed toy choices ranged from 0.54 to 1.92.[12]

At the age of two, the sexes differ in interest in toys, in aggressiveness, and in fearfulness.[13] Toddlers exhibit toy preferences matching familiar stereotypes.[14] Two-year-old boys tend to range farther in their play, engage in more outdoor play,[15] and display greater aggression than do girls.[16] The early appearance of these sex differences has led psychologist Anne Campbell and colleagues to suggest that "the neuroanatomical or neurochemical substrates of men's status-seeking may be laid down very early indeed."[17]

Although well-established sex differences in toy preference have a learned component, learning is not the whole story.[18] Children exhibit sex-typed preferences even before they can identify which toys are "sex-appropriate" for themselves.[19] By the time they reach three and four years of age children will justify their choices of toys for other children in terms of sex-role stereotypes, but they do not justify their own toy preferences in such terms. Rather, they justify their own preferences in terms of what the toy will do or some other characteristic of the toy.[20] This result is consistent with the finding that children are much less likely to stereotype themselves than they are to stereotype others of either the same or opposite sex.[21]

Same-sex playmate preferences are, as we saw in chapter 2, clearly exhibited by the third year in girls and the fourth year in boys, but these preferences often develop before children can reliably identify which children are the same sex as themselves. As Eleanor Maccoby has pointed out, an innate bias toward same-sex play is suggested by the fact that the same sex-segregation appears in nonhuman primates "among whom the cultural transmission of cognitive gender stereotypes is surely minimal."[22]

Social Influences

The three primary social forces acting directly on children are parents, teachers, and peers. In considering the influence of these agents, one should bear in mind that social influences might tend either to exaggerate or to minimize natural tendencies. There is reason to believe that in many cases—especially with respect to schools and teachers—there is more minimization than exaggeration.

Parents. Parents are often assumed to be the major culprits in sex-biased socialization. Although parents do sometimes provide differential feedback for certain kinds of behaviors, the scope of that differential reinforcement is less than many apparently believe.[23] For example, a meta-analysis of 172 studies examining differential socialization conducted by psychologists Hugh Lytton and David Romney found clear sex differences in encouragement of sex-typed activities and perceptions of sex-stereotyped characteristics, but this was the only one of eight major variables for which the researchers found differences. They also found that this differential treatment decreases as the children age. The other seven variables, for which clear differences were not found, were

1. amount of interaction;
2. total achievement encouragement (general and specifically with respect to mathematics);
3. warmth, nurturance, and responsiveness;
4. encouragement of dependency;
5. restrictiveness/low encouragement of independence;
6. disciplinary strictness; and
7. clarity of communication/use of reasoning.

They also found no support for the notion that with increased sensitivity to "sexist" rearing practices, parenting has become less sex-differentiated since the 1950s.

In their meta-analysis, Lytton and Romney cautioned against the simplistic view that the causal relationship between parental treatment and children's behavior runs only in a single direction. Rather, they noted, parental encouragement may build upon the child's already existing preferences; thus, different children may evoke different responses.[24] Although fathers are less likely to give dolls to boys than to girls, boys are also less likely than girls to play with dolls when given them. When parents purchase toys for their children that the children did not request, they are equally likely to choose sex-typed or non-sex-typed toys, while the toys they purchase upon request of the child are most likely to be sex-typed.[25] Similarly, the common finding that parents engage in more rough physical play with male infants than female infants seems to be a product of boys' greater receptivity to such stimulation.[26]

Failure to take into account the fact that parents respond to differences in their children leads to a vast overestimation of the impact of socialization. Psychologists Richard Bell and Lawrence Harper have sharply criticized the "decades of socialization research [that] have pursued a simple and plausible answer to the problems of human development—that most of the child's characteristics are brought about by the behavior of the parents."[27] Although the argument is plausible, it turns out to be wrong.[28] Thus, it must be understood that when Fausto-Sterling asserted that differences in genitalia cause parents to treat them differently and that this differential treatment in turn causes different behaviors in boys and girls, she is making a causal argument for which there is scant support. Similarly, Hennig and Jardim attribute the success of the executive women they studied to fathers who engaged in traditionally male activities with them.[29] It seems likely, however, that at least part of the reason that these fathers engaged in traditional male activities with them was that these particular girls, unlike many others, especially enjoyed these activities, possibly, as we will see in the next chapter, because they were exposed to higher-than-average levels of testosterone *in utero*.

Apparent sex differences in emotional needs have been attributed to the greater frequency with which parents caress and hold their infant daughters, thereby creating "in women a greater desire and need for emotional intimacy." [30] Again, this argument implies that parental behavior creates the need for emotional intimacy. Perhaps it is the other way around. A recent study showed that individuals blind to the sex of newborns rated female infants substantially more "cuddly" than male infants, with effect sizes of 0.69 and 0.80 for female and male experimenters, respectively.[31] This finding makes it problematic to conclude that later emotional sex differences were caused by the parents' differential cuddling of boys and girls; instead, it seems probable that parents are simply more likely to cuddle with particularly "cuddly" infants.

The pervasive assumption of unidirectional causation leads to frequent and often unrecognized assumptions about causation. In her study of play discussed in chapter 2, Janet Lever argued that encouragement of boys to engage in "boy" play is what turns them into leaders. She assumed that sex differences in play were caused by differential socialization: parents have encouraged contact sports for boys "because they believe the 'male nature' requires rough and tumble action, and organized competition is the best outlet for this surplus energy." On the other hand, she continued, "parents believe their girls are frail and less aggressive, and therefore do not enjoy serious competition; rather, they believe girls feel their maternal instincts early and prefer playing with dolls and reconstructing scenarios of the home." She further argued that the experience in competition and team sports that boys obtain in childhood may give them an advantage in their later work lives. Finally, in order to redress the disadvantage of girls, she suggested that physical education programs might be broadened "to include learning opportunities now found primarily in boys' play activities." [32]

Lever's educational prescription may be a good one, but it is not supported by the justifications she offered. Lever viewed the boys as being acted upon by their peer group and learning ways of thinking through games that are reflected in the later behavior of adult males. She did not advert to another possibility, one that does not indict the educational system: boys and girls played the games that they did because they were temperamentally suited to their respective games. In other words, boys simply enjoy vigorous competition and arguing over the rules more than girls do. Although Lever decried the absence of organized team sports for girls, there is no indication that any of the play she observed consisted of organized sports leagues. Instead, the games that she observed on the playground were spontaneously organized by the children themselves. Nonetheless, of all the many team games that she observed during a one-year period, only one was organized by girls.

Those who believe that children are shaped by their parents' stereotyped treatment should ask themselves whether they treat their own children differently because of their sex. It seems likely that they would say that they treat their own children as individuals, responding to their individual personalities and needs, and that it is only "other people" who regularly stereotype their children.

Schools. Schools are also often implicated in the sexist rearing of children. Teachers, it is argued, reinforce sex-specific behaviors in boys and girls. Again, for such a commonly accepted idea, there is remarkably little data to support it. A more accurate view is that teachers positively reinforce stereotypically feminine behaviors in both boys and girls and negatively reinforce masculine behaviors in both sexes.[33] As Warren Farrell has pointed out:

> From a boy's perspective, school itself is filled with women. It is women teaching him how to be a boy by conforming to what women tell him to do after he's been trained to conform to what his mother tells him to do. On the one hand, history books show him that his role is to be a hero who takes risks and, on the other, his female teacher is telling him not to take risks—to not roughhouse, not shout out an answer spontaneously, not use swear words, not refer to sex, not get his hair mussed, his clothes dirty . . .[34]

One researcher found in a study of sixth through eighth graders that the more feminine both boys and girls were, the higher were their grades in all subjects, irrespective of the sex of the teacher.[35]

The perception that schools are a major source of sexist socialization was substantially fueled by the release of the 1991 American Association of University Women report entitled *How Schools Shortchange Girls.* The report asserted that teachers lavished more attention on boys, calling on them more often and essentially ignoring the girls, leading to girls' loss of confidence and self-esteem. Although the report's conclusions were widely reported, there was little factual basis for the assertion of differential treatment by teachers.[36] While there is some basis for the claim that teachers are more likely to call on boys than girls, that disproportion is attributable to the fact that boys volunteer more often than girls.[37] Further, much of the attention that teachers direct toward boys is negative in character. Thus, the failure to appreciate the role of the child in adult-child interactions led to an unfounded—and, given the higher grades given by teachers to girls, highly implausible—accusation of sexism.

Contrary to what would be expected if teachers were agents of sexist socialization, sex segregation in schools is inversely proportional to the degree of control the teacher exercises in establishing structure. In classrooms, where teachers have the major role, there is much less segregation than on the playgrounds. When teachers allow a free choice of seats, however, boys sit with boys, and girls with girls. Thus, cultural forces are attempting to counter what seems to be a natural impulse to self-segregation rather than creating a socially driven segregation.

Peers. Perhaps the most powerful influence on children is their peers.[38] As described in chapter 2, once they begin to prefer same-sex playmates, children (especially boys) begin policing the "gender boundaries." By about age five, there is strong peer pressure on boys to avoid appearing a "sissy" or being identified with "girl things." The consistent finding that boys are more strictly confined by sex-role expectations than girls rests uneasily alongside the widespread view that sex-role expectations are especially damaging to girls. "Tomboy" and "sissy" are not equally pejorative terms; many women boast of having been tomboys as children, but few men boast of having been sissies.

Children's play behavior, which as we have seen is highly sex-typed, is not something that is directly taught by parents or teachers, nor is it learned by watching the behavior of parents and teachers. Rather, to the extent that it is learned, it is learned mostly in the culture of the child. This finding leads to the obvious question: why? Why do children so eagerly aim to enforce sex boundaries? Many teachers and parents consciously aim to provide a non-sex-typed environment, and most children in modern societies do not see their parents interacting in predominantly single-sex groups. Yet anecdotes abound of parents who were determined that their little boys not play with guns but who found that the boys would readily find acceptable substitutes, whether fingers, sticks, bananas, or even Barbie® dolls. Even parents who are relatively traditional themselves provide only limited sex-typed cues, such as fathers who strongly discourage their sons from playing with dolls. One must wonder how parents who consciously attempt to avoid sex-stereotyping still end up with children having the same basic ideas of sex roles.

Instinct versus Learning

The apparent readiness of children to assimilate sex roles and their attentiveness to members of their own sex requires some explanation. A purely sociocultural argument relying on the subtle but pervasive nature of societal cues begs the important question. Parents attempt by instruction and example to cause their children to clean their rooms, eat their vegetables, close the door, do their homework, and cooperate with their siblings. To the grief of many a parent, however, these lessons often do not easily take. Whence then comes the power of the lesson "act like a boy" or "act like a girl"? To the extent that learning is involved at all, the answer is most likely that children are biologically primed to internalize sex roles in much the same way they are primed to learn language.[39]

Evidence that a behavior is influenced by learning does not discredit a biological explanation. As psychologist Isaac Marks stated, "All species learn some things far more easily than they do others, a facility shaped by natural selection in particular environments."[40] Indeed, as Lionel Tiger has observed, one of the most fruitful routes toward understanding the psychology of humans or other animals is to ask what it is that is easy for them to learn.[41] Thus, when it is argued that children learn sex roles from watching other children, one should ask why sex roles are seemingly so easy for them to learn. Indeed, why will they learn these roles even if active measures are taken to prevent it? The answer seems to be that, in the words of psychologist Linda Mealey, "The learning of sex roles can be considered to be a kind of biologically prepared learning: boys and girls seem to be biologically predisposed to pay attention to gender differences in the behavior of others and to care about the gender appropriateness of their own behavior."[42]

Does Differential Treatment by Parents, Teachers, or Peers Have Any Effect?

When people assert that parents, teachers, and peers act in sex-typed ways toward a child, they generally assume that the behavior influences the child's behavior or personality. People assume that if parents say "don't be aggressive," then the child will be less aggressive; if parents say "be more aggressive," the child will be more aggressive. If these messages are given (or are thought to be given) differentially to boys and girls, that difference is then assumed to explain observed sex differences in aggressiveness. This view substantially overstates the effect of such influences and neglects the selectivity with which children respond to behavioral reinforcement.

As we have seen, what appears to be differential treatment based upon sex is often a reflection of preexisting temperamental traits of the children that evoke different parental behaviors. For those circumstances where differential treatment exists, however, the question remains how powerful the influence of parenting style is on children. The answer seems to be that it is surprisingly weak.[43] Parents consciously try to shape the values, interests, and personalities of their children but often without any discernible effect. Thus, even a clear showing of stereotypic treatment by parents would not demonstrate that it contributes to later characteristics of the child. As Morton Zuckerman put it, "all parents are environmentalists until they have their second child."[44] At that point, they tend quickly to abandon the notion that their children's personality was shaped by the home environment, because they see behavioral differences in their children that are too great to have been caused by any differential treatment they might have received.

The assumption that all reinforcement is equally effective underlies much of the writing in this area, but the reality is more complex, as children are quite selective in their responses to feedback. Their responses depend both upon the type of behavior involved and the identity of the person providing the feedback. Reinforcement for acting feminine affects girls more than boys, while reinforcement for acting masculine affects boys more.[45] Both sexes receive negative feedback for high activity from teachers, and while that feedback influences the girls it tends to be ignored by boys.[46] Similarly, both boys and girls tend to receive negative feedback from adults for aggression, but that feedback often tends to result in continuation or intensification of the aggression in boys.[47] Children (especially boys) are influenced more by same-sex feedback than they are by opposite-sex feedback.[48]

Many teachers and parents affirmatively attempt to eliminate sex typing, but these efforts appear to yield little in the way of behavioral change. Although explicitly teaching children that both sexes can perform particular jobs seems to decrease the extent to which children hold stereotyped views of the jobs, it does not affect the children's own highly sex-typed preferences.[49] In fact, sometimes "nonsexist intervention" in the schools can actually increase sex stereotyping, at least in boys.[50] Depicting adults in cross-sex occupations actually seems to increase the stereotyping of the children of the sex for whom the occupation is traditional.[51]

If the socialization explanation were a complete one, one might have thought that the dramatic changes in women's roles that have occurred in the last three decades—combined with conscious attempts by schools to avoid sex bias—would have had a substantial effect on children's behavior. However, there is little evidence that sex typing has become less pronounced, and traditional sex differences in play behavior have persisted.[52] Despite negative reinforcement, boys continue to exhibit boy-like behavior.[53] Similarly, despite changes in the roles of women and in people's attitudes about women, stereotypes of the two sexes have not become less differentiated over time.[54] According to social-role theory, stereotypes derive from the roles that society has "assigned" to the two sexes.[55] As these roles change, then stereotypes ought to change as well. However, despite substantial changes in what the sexes do, there seems to be little change in what people think that men and women are like.[56]

Implicit in social-role theory, at least in its strongest form, is the notion that if boys were raised as girls and girls as boys, then the psyches of the sexes would reverse. The story of "Joan/John" is perhaps the most well known refutation of that view. At eight months of age, the baby's penis was destroyed during circumcision. Despite being surgically reconstructed and raised as a girl, he rejected girls' toys, clothes, and activities and preferred those of boys; he even attempted to urinate standing up. Sometime between the ages of nine and eleven, he came to understand that he was not a girl. He was given hormones and reconstructive surgery and at age twenty-five married a woman.[57]

The strong sociocultural position views environmental influences on personality as virtually entirely determinative of later personality and preferences. This view seems to conflict, however, with the finding that personality tends to be relatively stable over decades. If personality were as sensitive to the vagaries of the environment as the social constructionists would have it, one would expect that aging, changes in social roles, and the occurrence of various life events would cause dramatic shifts in personality over time, but this tends not to be the case.[58] Instead, attributes of personality tend to be remarkably stable over time—timid children tend to become timid adults, extraverted children tend to become extraverted adults, and thrill-seeking children tend to become thrill-seeking adults (assuming they survive to do so). Thus, something more fundamental than culture-specific influences seems to affect the temperaments of the two sexes, and that something appears to be biology.

Cognitive Differences

The pure-socialization explanation for cognitive sex differences suffers from many of the flaws of the argument for temperamental differences. It asserts that boys have greater spatial skills because they are given more opportunities to develop them and that they have greater math skills because of the view that math is for boys and not for girls. In essence, girls and boys think differently because society expects, if not wants, them to.

Cognitive Sex Differences Appear Early and Cross-Culturally

Cognitive sex differences, like temperamental sex differences, are reported even in children, although the differences tend to be smaller in children and are often not detected when measured by tests developed for adults. In a study of nine- to thirteen-year-olds, Kimberly Kerns and Sheri Berenbaum found sex differences on a number of spatial tasks, with effect sizes of up to 0.65.[59] Sex differences in route learning that appear in adults have been replicated in children as young as five, with boys learning routes faster and girls recalling more landmarks.[60]

Some researchers believe that sex differences in spatial ability do not show up more consistently in children because the verbal advantage of young girls masks the male spatial advantage. Thus, while a number of studies have found that the male advantage in spatial performance does not reliably appear until age ten, a study controlling for verbal ability found a significant difference in spatial performance in first grade.[61] Indeed, as psychologist Diane Halpern notes, "the male advantage in transforming information in visual-spatial short-term memory is seen as early as it can be tested—perhaps at age three—and in mathematical giftedness as early as preschool."[62]

The sex differences observed in the United States are replicated cross-culturally, although the magnitude of the differences may vary.[63] The male advantage on mental rotation tasks found in the United States appears to be universal, having been found among Europeans, Africans, and Asians. In commenting on the findings of comparative studies, Diane Halpern has characterized them as revealing "amazing cross-cultural consistency."[64]

The male advantage in mathematics is widely replicated, but because mathematics must be studied in school it is more subject to cultural variation. Nonetheless, the Third International Mathematics and Science Study (TIMSS) confirmed the general pattern.[65] Of the twenty-one (mostly European) countries studied, boys' scores were significantly higher in eighteen. The three exceptions were South Africa, Hungary, and the United States, where the male advantage was not statistically significant for this particular test.

Girls in the United States are clearly not performing at the highest level that their native cognitive ability will permit, nor, of course, are boys. Among the twenty-one countries represented in the TIMSS, in fourteen the female score exceeded the score of U.S. males. Thus, neither boys nor girls in the United

States are in any danger of reaching a performance ceiling in the near future, a fact that has led to calls for improvement of math and science education.

Increasing the rigor of math and science education may have effects that are unwelcome to those who worry about sex differences. Sex differences on the TIMSS in mathematics general knowledge tend to be smallest in nations with the lowest average performance and largest in nations with the highest performance. Thus, South Africa, Hungary, and the United States—the three countries where the male superiority was not statistically significant—all performed below the international average. In fact, the lowest eight nations on general mathematics achievement (South Africa, Cyprus, the United States, the Czech Republic, Lithuania, the Russian Federation, Italy, and Hungary) include the seven nations with the smallest sex difference (the deviation from the pattern was the Czech Republic, which had both low average scores and a relatively large sexual disparity). On the other hand, the three nations with the largest sex difference in performance—Norway, the Netherlands, and Denmark—were among the six highest-scoring nations. They are also, of course, among the most sexually egalitarian countries in the survey, if not the world. Similarly, in science general knowledge, the four highest-scoring countries had sex gaps that were higher than average, and the six lowest-scoring countries had gaps that were smaller than average. Thus, it may well be that an effort to improve overall mathematical and scientific performance in the United States may have the effect of actually increasing sex differences.

The Limits of a Sociocultural Explanation for Cognitive Sex Differences

Those inclined toward purely social explanations argue that male spatial superiority is a consequence of greater experience with building blocks, video games, and throwing and other sports-related tasks. Male mathematical superiority, they say, is a consequence of the perception on the part of students, teachers, and parents that math and science are male domains. Boys are therefore encouraged to pursue such study and rewarded for doing so in a way that girls are not. There is remarkably little evidence that any of these phenomena, even if real, actually have a causal influence on anything.

Although it would be unwise to dismiss social contributors entirely, it seems unlikely that such factors as differential participation in sports or differential use of video games are complete or even important explanations for sex differences in spatial ability.[66] Sex differences are found even in preschoolers, although the magnitude is smaller,[67] and they appear even in countries that are viewed as more egalitarian than ours.[68] Although large sex differences in mathematics generally do not appear until the time of puberty, more boys than girls perform at the gifted level from preschool on.[69] However, Raymond and Benbow have found no evidence that intellectually talented males receive more encouragement in mathematics than talented females.[70]

Correlations between activities such as playing video games and spatial ability do not necessarily show that usage of video games enhances spatial ability; it is equally plausible that those with high spatial ability are drawn to video games, which reward that ability. To the extent that causation runs in that direction, studies that control for video-game usage will underestimate the sex difference. Moreover, while certain spatial skills can be improved through practice, males and females appear to benefit equally from additional practice,[71] implying that the average sex difference is not simply a reflection of differential experience. That is, if male superiority were a consequence of extensive practice, one would expect males to be much closer to their performance ceiling than females and therefore to benefit less from additional practice.

Some spatial tasks showing reliable sex differences are difficult to correlate with differential experience. The water-level task, for example, requires the subject to indicate where the water line would be on a tilted glass. The correct answer, of course, is a horizontal line, that is, a line that is parallel to the ground. Surprisingly, however, there are large sex differences in this task; indeed, approximately 50 percent of college women do not realize that the water remains horizontal.[72]

One often hears that high-ability girls face social pressures to "play dumb" and therefore fail to develop their full potential. Without question, such pressures exist, although it is not clear to what extent they actually affect academic performance. However, those pressures can explain sex differences in cognitive performance only if high-ability boys do not face and respond to the same pressures. In fact, they do. As Camilla Benbow and Julian Stanley have observed, high-ability students of both sexes "use denial, distraction (displaying excellence in another realm, preferably athletics), deviance (e.g., class clown), and underachievement."[73] It is also noteworthy that if girls, selectively, are "playing dumb," they are not doing it very well, since they get better grades than boys in virtually all classes, including math and science.

The argument that math is viewed as a male domain, resulting in reduced expectations on the part of students and teachers, is not intuitively appealing, despite its frequent repetition. If girls' performance is impeded because they believe that math is for boys, why do they do better than boys at arithmetic calculation? Why is the male advantage so selective—that is, primarily in geometry and algebraic word problems among secondary school students? As psychologist Diane McGuinness asks, "How could it be possible for society to encourage the aptitude of females in arithmetic, where they often show superior skills to males, but to discourage female performance in mathematical reasoning or higher mathematics?"[74] Moreover, she asks, "Why would society want girls to do badly in math, but not in other school subjects like foreign languages, art, and biology?"

It is not so clear that girls and boys actually think that math is for boys. One study asked mothers and children whether they thought boys or girls were better at math.[75] Almost two-thirds of mothers responded that they were the same, while the remainder were equally divided over whether boys or girls were superior. Among first graders, a majority of both boys and girls said that the sexes were equal, while most of the rest thought their own sex was better. Among fifth graders, a larger majority thought the sexes were equal, and the tendency to view their own sex as superior had increased. Only 6 percent of the girls thought boys were better, and 10 percent of boys thought that girls were better. While one cannot entirely rule out the possibility that respondents were giving socially desirable answers, these results do not suggest a pervasive belief among any group that math is for boys.

If girls are trying (either consciously or subconsciously) to hide their innate mathematical ability, one would expect them to do relatively better on anonymous standardized tests than they do on more visible tasks, such as classroom performance. Yet, the observed pattern is just the opposite. Girls get better grades in school than boys do, even in math, but do less well on standardized tests of conceptual ability.[76] Moreover, if teachers believe that boys are better at math than girls, it is odd that they keep giving girls better grades.[77]

Some believe that a social explanation is rendered more likely by the fact that the largest mathematical sex differences do not appear until around the onset of puberty. This argument rests on the implicit assumption that features that are not apparent at birth are socially caused. But we do not believe that menstruation, pubic hair, and breast development are caused by society just because they do not occur until puberty. Moreover, the late onset may have less to do with changes in boys and girls than it does with changes with age in what is measured by the various tests. Diane Halpern has pointed out that the tasks showing the largest sex differences in math require visual-spatial processing of information—"tasks such as geometry, calculus, and converting the information in a word problem to a spatial representation."[78] These skills are not required in elementary school, and in those grades there tends to be no sex differences or a female advantage. Only after the curriculum becomes complicated enough to require these higher-order skills do the strongest sex differences emerge.

The Kibbutz: A Laboratory Experiment?

If we could design an experiment to test the influence of socialization, we might construct a society built upon an ideology of sexual equality, where sex roles are eliminated and women are liberated from the burden of domestic obligations. The Israeli kibbutzim provided just such a test, and it is not kind to the pure-socialization argument.

The kibbutz movement, which began in the early part of the twentieth century, was based upon a sexually egalitarian ideology. The kibbutzim were established with the specific goal of redressing the "biological tragedy of women," which caused women to be dependent upon men and limited to the domestic sphere. In the kibbutz, children no longer lived with their parents but in children's houses with age mates. Communal kitchens, dining rooms, and laundries were created to free women from domestic obligations and allow them to work side by side with men. Men and women were expected to share equally in political power. Women shunned feminine dress and jewelry.

Two to three generations later, the position of women in the kibbutz was the subject of two extensive studies, one by anthropologists Lionel Tiger and Joseph Shepher and one by sociologist Melford Spiro. As they both reported, life turned out very differently from the way it was conceived by the founders. The strong version of the founding generation's ideology lasted for about one generation. Although in the early years, women worked side by side with men, by the 1950s a high degree of occupational segregation had reemerged, and men were doing the high-status farming, and women were working as teachers and nurses. Women continued to do the laundry and cooking, but in their service-sector jobs rather than at home. Men held most of the positions of authority, and the higher the position, the stronger this tendency. Although a one-third quota for women was established for governing bodies in one kibbutz federation, the quota could not be satisfied because not enough women were willing to serve.

Subsequent generations found little to emulate in the founders' androgyny. Whereas women of the founding generation sought to minimize sexual differences in clothing, their granddaughters found a revived interest in fashion and jewelry. Whereas women of the founding generation felt relief at not having to live with their children, their granddaughters (and to a lesser extent, their grandsons) felt an emptiness that led to pressure to eliminate the children's houses. As a result, the family rose from its initial "shadowy existence" to become the basic unit of kibbutz social structure.[79] Life in the kibbutz has changed so much that sex-role distinctions are now greater in the kibbutz than outside.[80]

The kibbutz experiment suggests that the sex roles assumed to be mere cultural artifacts have deeper origins. Melford Spiro describes his study of the kibbutz as forcing upon him "a kind of Copernican revolution" in his thinking: "As a cultural determinist, my aim . . . was to observe the influence of culture on human nature. . . . I found (against my own intentions) that I was observing the influence of human nature on culture."[81] The reversion to more traditional sex roles in the kibbutz did not reflect a rejection of the idea of sexual equality; that ideology endured. Rather, it represented a rejection of the particular version of sexual equality that entails sexual sameness, coupled with a recognition on the part of the members—especially women—that traditional roles can be more fulfilling than the roles impressed upon them by the ideology of their grandparents.

If socialization is an inadequate explanation, what is to take its place? We will see that the answer is biology—not a crude biological determinism that sees humans as puppets on the strings of their genes, but a biology that creates predispositions in individuals that incline them more strongly in some directions than in others in ways that are responsive to pressures existing in our evolutionary environment.

8 Hormones

The Proximate Cause of Physical and Psychological Sexual Dimorphism

Sex differences described in prior chapters are created both by organizing effects of sex hormones—permanent effects on brain development that occur primarily in the womb—and activational effects—more transitory effects resulting from the immediate influence of circulating hormones.

The process of sexual differentiation is critically dependent upon release by the testes of testosterone during fetal development, without which the body and the brain develop in the female direction. Females exposed to higher-than-usual levels of testosterone in utero tend to be somewhat masculinized temperamentally, exhibiting more male-like behavior patterns and occupational preferences, while males deprived of the effects of testosterone in utero develop in a more feminine direction. Levels of circulating testosterone also appear to affect behavior, principally dominance behavior.

Sex hormones similarly have organizing and activational effects on cognitive ability. Girls exposed to high levels of testosterone in utero have higher-than-average spatial ability and boys exposed to abnormally low levels have lower-than-average ability. Spatial ability varies with levels of circulating hormones as well, with the optimal level apparently in the low-normal male range. Thus, men with relatively low testosterone and women with relatively high testosterone tend to have the highest performance.

The biological argument for psychological sex differences requires us to start at the very beginning and examine what it is exactly that makes a male and a female. The process of sexual differentiation is complex and affects not only the body of the developing human but also its brain.

For the first two months or so after conception, the process of growth and development is quite similar for embryos with XX (female) and XY (male) chromosomal complements.[1] The primordial genital tract of both sexes is undifferentiated. If the embryo is XY, differentiation in the male direction is initiated by a gene on the Y chromosome, the *Sry* gene (sex-determining region of the Y chromosome), which causes the bipotential gonads to develop into testes.[2] The testes, in turn, produce substances that cause the masculinization of the reproductive system. If masculinization does not occur during the critical period of development, the fetus will develop as a female; it is for this reason that the female body form is sometimes called the "default" form.[3] Thus, as is the general mammalian pattern, once the sex of the gonads is determined, human sexual differentiation occurs primarily through the actions of gonadal hormones.[4]

Although androgens—such as testosterone and dihydrotestosterone—are commonly referred to as *male* sex hormones, and estrogens—such as estradiol—and progestogens—such as progesterone—are commonly referred to as *female* sex hormones, both classes of hormones exist in both sexes. The important developmental differences between the sexes are therefore not a consequence of completely different hormones, but rather of differences in hormone levels and timing of exposure.[5]

The sexual differentiation of the brain, like that of the body, is critically dependent upon hormones. During a critical period of development during the second trimester, androgens cause the male brain to become masculinized.[6] As with anatomical development, the determinant is hormonal exposure. Unlike the situation with genital differentiation, however, there is some evidence that prenatal estrogens play an active role in brain differentiation.[7]

If a chromosomal female is exposed to high levels of androgens during the critical period, she will be psychologically masculinized. Likewise, if a chromosomal male is effectively deprived of a sufficient level of androgens either because they are not present or are not biologically active, he will be psychologically feminized. These developmental hormonal effects are called "organizing" effects, in contrast to "activational" effects, which are the result of the immediate influence of circulating hormones on behavior. Although this clear dichotomy between organizing and activating effects may be oversimplified,[8] it is useful for discussion.

Hormonal Effects on Temperament

Evidence for hormonal influences on temperament comes from a variety of sources. Fetuses sometimes receive abnormal hormonal exposure because of a variety of genetic abnormalities, and sometimes they are exposed to abnormal hormones as a consequence of hormones administered to their pregnant mothers. Moreover, levels of the mother's own naturally occurring hormones may influence development of her child.

Congenital Adrenal Hyperplasia

The best-studied condition in which females have been exposed to high levels of male hormones in utero is *congenital adrenal hyperplasia* (CAH). CAH is a genetic anomaly that results from a defect in the synthesis of cortisol, an adrenal hormone. Because levels of cortisol are too low, the feedback mechanism between the pituitary and the adrenal gland causes increased adrenal hormone production, including production of adrenal androgens.[9] While both boys and girls can have this syndrome, the excessive adrenal androgen seems to have relatively little effect on boys, who are already exposed to high levels of androgens from their testes. Girls, however, can be substantially affected.[10]

The exposure of the female CAH fetus to androgens comes too late to cause masculinization of the internal reproductive system, but it does cause varying degrees of masculinization of the external genitalia. The condition is generally diagnosed at, or soon after, birth. The cortisol deficiency is remedied through supplementation, thus lowering adrenal androgen levels to normal, and the genitalia are surgically corrected. These girls generally develop as normal fertile females, although the onset of menstruation is often delayed.

The behavioral profile of CAH girls as children is strikingly more masculine than unaffected girls.[11] They are much more likely to be "tomboys" throughout childhood. They tend to reject feminine fashions and "girl toys" such as dolls, preferring functional clothing and "boy toys," and they are considerably more likely than controls to choose boys as playmates.[12] CAH females, as adolescents and adults, are more likely than unaffected females to use physical aggression in conflicts.[13] They also exhibit substantially less interest in infants and express a lesser desire to marry and have children when they grow up,[14] and they have more male-like occupational preferences.[15] They are also less likely to be classified as exclusively heterosexual and may have an unstable gender identity.[16]

The CAH evidence is highly suggestive of hormonal effects on psychosexual differentiation. Still, it must be viewed with some caution, because the population is an abnormal one. Not only do the girls receive supplemental cortisol, but, perhaps even more important, the parents know that the girls were born with abnormal genitalia. Because of the possibility that parents may be more accepting of male-type behavior in CAH girls than they would be in unaffected girls, some have argued that no inferences at all can be drawn from these girls.[17] There is little empirical support for the proposition that parents of CAH girls treat them differently, however.[18] Indeed, some researchers have argued that parents' knowledge of their daughters' abnormal exposure would, if anything, incline them to be even less accepting of male-like behavior from their daughters.[19] Moreover, if genital virilization results in differential parental attitudes, one would expect that the greater the degree of genital virilization, the greater the degree of parental acceptance of masculine behavior, but no such relationship has been found.[20]

Sheri Berenbaum recently tested indirectly the hypothesis that parental treatment is responsible for the masculine tendencies of CAH girls.[21] She reasoned that if genital virilization affects the way people interact with CAH girls, then CAH girls' preferences and behavior should become more female-typical with age. In the early years, a large part of the girls' potentially influential environment—their parents—would be aware of their genital abnormalities and might treat them accordingly. However, as children age, an increasingly large portion of their social environment consists of schools and peer groups. Teachers and peers would generally be unaware of the girls' abnormal genitals, which would have been corrected by surgery. Thus, as the girls age, their environment would become increasingly female-typical and their behavior should become so as well. If behavioral masculinization results from permanent organizing effects in utero, however, the girls' masculine tendencies should continue into adolescence.

Berenbaum measured participation in sex-typed activities and interest in sex-typed occupations and found that adolescent CAH girls differed substantially from control girls on all six of the scales measured, although one difference, though not insubstantial in size ($d = 0.4$), was not statistically significant. The rest of the differences were both significant and quite large (d's ranging from 0.8 to 1.4). The scores of CAH girls also differed significantly from those of control boys. The intermediate position of the CAH girls meant that unlike either set of controls, they had no significant sex-typed activity and interest preferences. These results strongly suggest that the masculinization of CAH girls' behavior and preferences are caused by organizing effects of androgens rather than by social responses to the girls' genitalia.

Other Hormonal Evidence

Corroborative evidence for the hormonal hypothesis comes from individuals with *Androgen Insensitivity Syndrome* (AIS). AIS individuals have the XY chromosomal complement of a male, and their testes develop and produce testosterone, but their tissues lack androgen receptors.[22] AIS patients tend to exhibit stereotypically female preferences (such as a desire to be a wife with no outside job) and tend to be interested in infants and dolls. Interpretation of these findings is complicated, of course, by the fact that the subjects are also reared as girls.

Another informative condition is *5-α-reductase pseudohermaphroditism*. This condition occurs when an XY male has functioning testes, adequate androgen receptors, but a deficiency in the enzyme 5-α-reductase. The testes produce testosterone, but the enzymatic deficiency prevents the conversion of testosterone to dihydrotestosterone (DHT), which is critical to the development of the external genitalia. Babies born with this deficiency appear to be girls, but the flood of DHT accompanying puberty causes male sexual development, so these children appear to change sex at puberty. They are mostly able to assume the male sex role without substantial difficulty, leading investigators to conclude that socialization has little impact on adult gender identity.[23] Because the condition is now well known in the areas in which it appears, however, parents generally know the status of their children, so the children's socialization may be substantially affected. Nonetheless, it does not appear that even the earliest cases revealed substantial difficulty, suggesting that knowledge of the effects of the condition may not be responsible for the ease with which the patients adjust.[24]

In addition to "experiments of nature" like CAH, AIS, and 5-α-reductase deficiency, data on prenatal hormonal exposure also come from studies of offspring born to mothers treated with hormones for maintenance of high-risk pregnancy. Although the findings are complicated by the fact that a variety of hormones are administered, sometimes in combination, a comprehensive review concluded that some of the hormones had masculinizing (or defeminizing) effects, while others had feminizing (or demasculinizing) effects. The behaviors that appeared most affected by administration of hormones were play, aggression and assertion, and gender identity/role.[25]

The circulating testosterone of the mother is another source of hormones for the fetus. The mother's testosterone levels during the second trimester of pregnancy are negatively related to the extent of her daughter's sex-typed behavior decades later in adulthood. Indeed, the mother's testosterone level during pregnancy seems to be a better predictor of her daughter's behavior than the daughter's own adult testosterone level.[26] Confounding those who would make an absolute separation between socialization and biology, sociologist J. Richard Udry has found that females' responsiveness to encouragement of femininity is inversely related to their mothers' testosterone levels during the second trimester of pregnancy.[27] That is, females exposed to high levels in utero seem to be "immunized" against the effects of feminine socialization; no matter how much encouragement their mother provides for feminine behavior, it has little effect and the daughter remains relatively masculine. Females exposed to low levels of prenatal testosterone, on the other hand, are more variable in their femininity, depending upon the extent to which they have been socialized in a feminine direction.

Animal Studies

Only in animal studies are properly controlled experiments possible, and they reveal compelling evidence for an organizing effect of hormones in all of the mammalian species studied. For example, female rhesus monkeys injected with androgens during the prenatal critical period later display stereotypic male sexual and nonsexual behavior. They engage in higher frequencies of threat behavior, play initiation, rough-and-tumble play, and chasing play. Females who receive testosterone throughout gestation have masculinized genitalia and behavior.[28]

Each of the above sources of information is suggestive of an organizing effect in humans, but perhaps not compelling standing alone. Taken together, however, a consistent relationship between hormones and behavior emerges that is not easy to dismiss by invoking different ad hoc criticisms of the individual findings. While we might wish for the kind of data in humans that we could get from the well-controlled experimental manipulation of hormonal levels in animals, few of us would wish to live in a society in which that was possible. Those who are determined to reject biological explanations because perfectly controlled biological experiments are not possible should perhaps reconsider their acceptance of sociological explanations given the similar impossibility of perfectly controlled sociological studies.

The Activational Effect of Circulating Hormones

Sex hormones also appear to have activational effects on temperament. For a number of reasons, data concerning the effects of circulating hormones are often ambiguous. The strength of the activational effect is probably substantially dependent on the extent of prior organizing influences.[29] Thus, a given level of testosterone, for example, would have different effects on the behavior of different individuals depending upon the extent to which they had been exposed to testosterone in utero.[30] This differential sensitivity would weaken any measurable correlation between hormone levels and behavior.[31] Moreover, hormone levels are constantly changing in regular daily and annual cycles (testosterone in men and women) and monthly cycles (estrogen in women), and also in response to social influences. As a result, it is difficult to identify the precise time that hormone levels should be assessed for a given comparison.[32] Finally, the lag time between release of the hormone and induced gene effects means that hormone levels in blood and in target tissues do not always correlate directly with their measurable effects.[33]

Probably the most studied effect of circulating androgens relates to aggression and dominance behavior.[34] Prison inmates convicted of violent crimes, for example, have higher concentrations of salivary testosterone than those convicted of nonviolent crimes, and once in prison, prisoners with higher levels of testosterone engage in more rule violations.[35] A significant relationship has also been found between testosterone levels and criminal convictions.[36] Trial lawyers have been found to have higher levels of testosterone than other lawyers, and female lawyers have higher levels than housewives.[37]

Sociologists Allan Mazur and Alan Booth have recently argued that testosterone is more directly related to dominance behavior than it is to aggression.[38] They contend that the reason that associations between testosterone and aggression are inconsistent is that only in some circumstances will dominance behavior take an aggressive form. Thus, they argue, prison inmates with high testosterone engage in high levels of physical aggression because prisons are social environments in which aggressive forms of dominance behavior are socially sanctioned (among the prisoners). While physical aggression may lead to dominance in a prison or a street gang, it will not do so in the ordinary workplace. A number of other studies have found a positive relationship between testosterone levels and social dominance.[39]

Testosterone thus appears to have a role in both encouraging competition for status in the first place and in influencing how men react to the results of competition. This competition need not be aggressive. As psychologist Dov Cohen has observed, "Testosterone may facilitate successful boardroom maneuvering as much as successful barroom brawling, or it may produce useful creativity as well as antisocial rebellion."[40]

Most of the studies of activational effects are merely correlational, so one must be cautious about inferring causation. A correlation might exist because either (or both) of the two variables might be the cause or both variables may be effects of some other cause. Thus, the question is whether the hormone levels caused the dominance or aggressive behavior or whether the behaviors caused the hormone levels. Although the answer to both questions is probably in the affirmative, the evidence of behavioral and situational effects on hormone levels is actually greater at this time than the evidence for the effects of hormones on behavior.

Most studies of the relationship between testosterone and behavior have focused on men. The pattern for women seems somewhat different. There is some evidence that androgen levels are related to assertive behavior in women, although the link to status is less clear. Indeed, Elizabeth Cashdan has found that high testosterone is negatively associated with status within female groups.[41]

In nonhuman animals, even in birds, the activational effect of androgens is clear.[42] The place of a hen in the "pecking order" can be increased by testosterone injections.[43] In mammals, administration of testosterone to female guinea pigs has been shown to result in the display of male-like sexual behavior.[44] Similarly, injections of testosterone into female rhesus monkeys not only increase their aggressive behavior but also their dominance status.[45] Testosterone injections reduce nurturant behavior in a variety of species.[46]

Hormones and Cognitive Ability

Sex hormones contribute not only to temperamental differences but also to cognitive differences. Here, also, they appear to have both organizing and activational effects. Much of the evidence for organizing effects on human cognition, as with temperament, comes from studies of girls with congenital adrenal hyperplasia. The most commonly replicated result in cognitive studies is that CAH girls substantially exceed control girls in spatial ability, especially on those spatial tasks showing the greatest sex difference in normal subjects. For example, Susan Resnick and her colleagues found an effect size between CAH females and control females of approximately 0.93 and 0.80 on two mental rotation tasks. No significant differences were found between male CAH patients and controls. Similarly, Elizabeth Hampson and her colleagues in a study of pre-adolescent children found that CAH girls exhibited higher spatial performance than controls.[47] A recent study found that drawings by kindergarten girls with CAH were similar to drawings by boys—with a characteristic emphasis on moving objects and cold colors—but quite dissimilar to those by unaffected girls—which are much more likely to include people and flowers and be rendered in warm colors.[48]

Not all evidence of a relationship between prenatal testosterone levels and later spatial performance comes from abnormal populations, however. The spatial ability of seven-year-old girls, for example, has been found to be positively correlated with prenatal testosterone levels as assessed in second trimester amniotic fluid.[49]

Sex hormones also appear to exert an activational effect on cognitive abilities. Cathleen Gouchie and Doreen Kimura have suggested that the relationship between spatial ability and testosterone levels is described by an inverted U-shaped curve.[50] That is, spatial ability is lowest in those with the very lowest and the very highest testosterone levels, with the optimal testosterone level lying in the lower end of the normal male range. Thus, males with testosterone in the low-normal range have the highest spatial ability among males, while females with relatively high levels (for females) show the greatest ability.[51] After the testosterone levels of older men have declined substantially (along with their spatial performance), administration of testosterone enhances spatial cognition but not other cognitive functions.[52]

Another indicator of androgen effects is the finding by behavioral scientist Patricia Hausman that women who have high levels of both mathematical and mechanical ability—a pattern rare in women but much more typical of men—exhibit a number of traits suggesting high levels of androgens (or low levels of estrogen). These females are taller, thinner, and older at menarche. Especially significant, perhaps, is the fact that these women are at a very high risk of losing a pregnancy.[53]

Sex differences in spatial ability do not seem to be solely a consequence of the enhancing effects of testosterone, as estrogen appears to exert a depressing effect. Indeed, a substantial portion of the increased sex differences in spatial ability observed after puberty seems to be a result of the depressing effect of estrogens. It has been suggested that this depressing effect is responsible for the tendency of distinctively feminized women to exhibit relatively low spatial ability.[54]

Fluctuations in cognitive performance are associated with temporal hormonal fluctuations. Female performance on cognitive tasks varies depending upon the phase of the menstrual cycle. Spatial performance tends to be highest in those phases of the cycle when estrogen levels are low (and therefore the testosterone/estrogen ratio is at its highest), and performance on verbal tasks that show a female advantage tends to be highest in the high-estrogen portions of the cycle.[55] Levels of testosterone in males also fluctuate; there is a daily cycle and an annual cycle, with levels being greater in the morning than in the afternoon and greater in the fall than in the spring.[56] Consistent with Gouchie and Kimura's finding that low-normal testosterone levels are optimal in males, male spatial performance is greater in the afternoon than in the morning and greater in the spring than in the fall. None of these periodic fluctuations comes close to eliminating the sex difference in spatial performance, however.

Although low-normal levels of testosterone are associated with high spatial ability in men, abnormally low testosterone levels are associated with low spatial ability (though normal verbal ability).[57]

Confirmatory evidence of a hormonal contribution to cognitive abilities comes from studies of transsexuals. Transsexuals are individuals whose gender identity—that is, the sex to which they subjectively feel they belong—does not match the sex of their anatomy. Surgery to conform the genitals to the "brain sex" is used in combination with hormonal treatments to "reassign" males to the female sex or vice versa. Male-to-female (MF) transsexuals receive estrogen, as well as anti-androgens, which block the effects of androgens, while female-to-male (FM) transsexuals receive androgen therapy. Prior to hormone treatment, hormonal levels of MFs are within the normal male range, and those of FMs are within the normal female range. After three months of hormonal therapy, however, hormonal levels have reached those of the target sex.[58] By that time, the FMs have experienced a clear and substantial increase in spatial ability. Interestingly, after having received testosterone injections for approximately one year, FMs showed no decline in spatial performance during the five-week period that hormone treatment was suspended for the sex-reassignment surgery, suggesting the possibility that testosterone has organizing effects even in adults. A study of verbal reasoning in MF transsexuals revealed cognitive effects of estrogens, with MFs experiencing enhanced verbal memory performance.[59]

Studies on animals have yielded results consistent with the human studies. In laboratory rats, for example, males generally outperform females on the maze task, a measure of spatial ability. If female rats are given androgens during critical periods of development, however, they perform as well as males and better than males castrated prior to the critical period.[60] Rhesus monkeys have also been shown to exhibit sex differences in spatial ability and, as in humans, to show diminished sex differences with old age.[61] Whatever sociocultural factors to which one might attribute sex differences in humans—video games, social expectations, and so forth—would seem to lack utility in explaining the findings in monkeys.

The hormonal literature paints a compelling picture for both activational and organizing effects. The hormonal findings demonstrate that men are not more aggressive or more spatial because they receive some specific aggressiveness or spatial-ability gene that women do not have. Instead, the expression of identical genes may be different in different hormonal environments. Of course, these differing hormonal environments themselves have a genetic cause, the *Sry* gene on the Y chromosome, which causes development of testes.

Now that we see *how* the sexes come to differ in temperament and cognitive ability, it is time to examine *why* they do so. That is, why is our species (as well as so many others) "designed" in such a way that the sexes differ in these dimensions? The answer is evolution through natural selection, which is the topic of the next chapter.

9 Evolutionary Theory and the Ultimate Cause of Biological Sex Differences

If women didn't exist, all the money in the world would have no meaning.
—Aristotle Onassis

Human sex differences are products of the same selective forces that cause behavioral sex differences in other species. The greater minimum parental investment of females that is required by mammalian reproduction makes females a resource for which males compete. Because male reproductive success is more variable than that of females, the reproductive gains to male-male competition can be large. Across human history, male status has led to greater reproductive success, leading to a predisposition among males to engage in the kinds of status competition that today so often have workplace implications.

Just as women cross-culturally are attracted to high-status males, men are attracted to women of high reproductive value—young, attractive, and healthy. Because women, unlike males, cannot generally enhance their reproductive success by acquiring status or accumulating mates, they have increased their reproductive success primarily through direct investment in, and care of, offspring.

The greater female parental investment also explains men's greater pursuit of short-term sexual strategies and women's greater discrimination in sex partners. An imprudent mating imposes much greater costs, on average, to women than to men, and a man may substantially increase his lifetime reproductive success through little more investment than a brief dalliance. Crudely speaking, men have a greater interest in mate quantity, and women in mate quality.

Sex differences in cognitive abilities appear also to be products of selective pressures. Enhanced spatial ability would have been a substantial advantage to hunters bringing down moving prey with projectile weapons, as well as to hunters returning to camp without needing to retrace their steps. In contrast, object location would be useful to a gatherer repeatedly returning to a reliable source of food. Mathematical ability, in contrast, would not have been selected for directly, but it may be a byproduct of enhanced spatial ability.

The place of humans in the animal kingdom is well established, and our firm rooting in the animal kingdom should make it apparent that our species has been shaped by the same forces that shaped our primate relatives. Despite the popular tendency to envision a clear break between animals and humans—with animal behavior being viewed as largely fixed by biology and human behavior viewed as largely independent of it—many students of behavior now reject that sharp dichotomy, believing animal behavior to be more environmentally sensitive and human behavior more biologically influenced than previously assumed.[1] This learning has so far failed to be assimilated by many of those studying human behavior—a failure labeled *anthropodenial* by primatologist Frans de Waal[2]—and the failure has been even more complete among those attempting to shape public policy.

The Logic of Natural and Sexual Selection

The theory of natural selection proffered by Charles Darwin almost a century and a half ago continues to occupy a central position in evolutionary biology.[3] Modern biologists adhere to Darwin's explanation that the necessary components of evolution by natural selection are *variation* (organisms within a species must vary with respect to the trait in question); *heritability* (the trait must be capable of being passed on to the organism's offspring); and *differential reproductive success* (some individuals must leave behind more offspring than others). The logic of natural selection is elegant in its simplicity: if those organisms possessing a certain genetic variant systematically leave more offspring than their conspecifics, there will be a corresponding change in the genetic makeup of the species with the favorable variants becoming increasingly prevalent.

Although in the popular view natural selection is understood in terms of differential mortality ("survival of the fittest"), differential mortality is important only insofar as it is reflected in differential reproductive success. The salmon who gives its last bit of energy swimming upstream to spawn and die is a success, while a salmon who cavorts in the ocean for several years may live a long and happy (for a salmon) life, but a genetic predisposition toward celibate frolic in the ocean will die out along with its bearer.

Some selective pressures operate similarly on the two sexes in virtually all species. Darwin's "hostile forces of nature" place pressures that are identical, or nearly so, on male and female snakes, monkeys, and humans. They all must secure adequate nutrition, they must be protected against too much and too little water and heat, and they all require protection from predators.

Other pressures—primarily those related to reproduction—may operate quite differently on the two sexes, resulting in either physical or behavioral dimorphism. Thus, peacocks have brightly colored tails that are attractive to peahens, and male bowerbirds build complex edifices to woo females. Neither a physiologically expensive and physically handicapping tail nor construction of magnificent bowers contributes to these males' survival, but both provide a basis for the exercise of female choice. Similarly, neither the grossly larger size of male elephant seals nor the antlers of many species of deer are particularly important for survival, but they both provide the means by which males compete among themselves for access to fertile females.[4] Darwin called selection based upon mating success "sexual selection," to contrast it with "natural" selection,[5] but as philosopher Helena Cronin has observed, "for modern Darwinism, nothing remains of the traditional idea that the intraspecific and social nature of sexual selection sets it apart from natural selection."[6]

Relative Parental Investment

Sexual selection is of central concern to this book, because the differential experience of men and women in the workplace is, in important ways, a reflection of the different selective forces that have operated upon human (and pre-human) males and females. That is, sexual selection has left humans with a sexually dimorphic mind that inclines men and women in somewhat different directions.

The difference in "reproductive strategies" of males and females is central to our discussion, although the term *strategy* should not be taken to mean a conscious process, but rather a behavior pattern that would be adopted if a conscious design were at work. Reproductive strategies involve the questions of when and with whom to mate, and they may differ between the sexes in how much of their reproductive effort is "mating effort" and how much is "parental effort."

The divergent reproductive strategies of males and females date back to the dawn of sexual reproduction. Indeed, the very definition of "male" and "female" foreshadows the differing reproductive interests of the two sexes. The sex that produces the smaller, usually more mobile, gamete is labeled "male," and the sex that produces the larger, more nutrient-rich, and usually less mobile, gamete is "female." This asymmetry in gametes presages later asymmetry in other measures of investment in offspring.

Biologist Robert Trivers has demonstrated that the force governing sexual selection is the relative parental investment of the sexes in their offspring. Trivers defined parental investment as "any investment by the parent in an individual offspring that increases the offspring's chance of surviving (and hence reproductive success) at the cost of the parent's ability to invest in other offspring."[7] Trivers showed that the sex with the greater typical parental investment will become the limiting resource, and individuals of the sex investing less will compete among themselves to mate with members of the more heavily investing sex. Members of the sex investing less can increase their reproductive success through numerous partners in a way that members of the other sex cannot. Trivers's predictions have been shown to be correct for a wide variety of animals. In most animals, the lack of male parental investment leads males to compete among themselves either through female choice or male-male competition.

Humans differ from over 95 percent of mammalian species in that males provide substantial parental investment.[8] A large sex difference in minimum necessary investment nonetheless remains, however. The combination of substantial paternal investment and a large remaining differential creates substantially different reproductive strategies for human males and females. The act of intercourse requires the investment of a few minutes of time for both the male and the female partner. If conception follows, the ensuing burdens on the two partners are grossly asymmetrical. The nine months of gestation can be provided only by the woman, and throughout most of our evolutionary history the baby was thereafter dependent upon its mother for milk. During gestation, the mother cannot become pregnant again and therefore cannot directly enhance her reproductive success through additional matings. Furthermore, during lactation, which in traditional societies may last three to four years, pregnancy is much less likely and might cause a reduction in the mother's investment in the first child, decreasing its likelihood of survival.[9] The father, in contrast, can pursue other reproductive opportunities while the mother is pregnant or nursing. However, if he is to invest in his children, certainty of paternity becomes very important, as a man who invests in children not his own may substantially impair his reproductive success.

The ability of males to enhance their reproductive success through multiple matings has far-reaching consequences. If some men are able to monopolize the reproductive potential of multiple women, other men are likely to be excluded from reproduction altogether. Indeed, in most societies, unmated men outnumber unmated women.[10] So, while many more males than females will never have any offspring, the most successful males will have many more offspring than the most successful female. Thus, men must compete with one another for access to females, with the losers in the competition facing extinction of their genetic line. The greater male than female variance in reproductive success means that the stakes of the mating game are higher for males, making it critically important to males to develop the attributes that lead to reproductive success. As we have seen, some of those attributes have important workplace implications.

Parental-investment theory provides an explanation for the well-documented differences in sexuality discussed in chapter 2. As psychologists David Buss and David Schmitt have shown, short-term sexual strategies have been, and continue to be, a greater part of men's repertoire than of women's. Historically men have increased their reproductive success primarily by increasing the number of their sexual partners rather than through an increase in the number of offspring per partner. A married man with two children could potentially increase his reproductive success by 50 percent through a single copulation that resulted in the birth of an offspring that survived to reproduce.[11] Females stand to benefit reproductively somewhat less through short-term strategies. Women cannot directly increase their reproductive success through multiple matings in the same way that men can, as a woman who has sexual relations with ten men over a short period of time will probably end up with, at most, one child (and, because of men's reluctance to invest in children not their own, likely no mate to provide support).

The costs of short-term mating also vary by sex. Because of asymmetry in parental certainty, a woman's promiscuity is a potentially greater threat to her long-term mate value than is a man's. Men are often unwilling to invest substantially in a woman who has shown herself willing to have sexual relations with many men, at least in cultural contexts in which premarital sex is a negative indicator of what investing men are critically concerned with—fidelity after marriage. Moreover, sexual promiscuity in a female may create the appearance that she is unable to attract a long-term mate and therefore must settle for short-term mating,[12] while for a man it may be viewed at least somewhat positively, indicating his desirability as a mate.

Status and Male Reproductive Success

In traditional societies, status and access to resources are among the strongest predictors of a man's reproductive success, while youth and beauty are among the traits of the most desirable women. Laura Betzig, in a far-reaching survey of status and reproductive success, found a high correlation between a man's power and his access to women.[13] Anthropologist Edgar Gregersen concluded from his study of three hundred mostly nonurban non-Western cultures that "for women the world over, male attractiveness is bound up with social status, or skills, strength, bravery, prowess, and similar qualities" and that "men are usually aroused more than women by physical appearance."[14]

The greater competitiveness and risk taking of men can be understood only in terms of the reproductive rewards that have, across time, accrued to high-status men. The reproductive payoff that comes from achievement of status has left them more inclined than women to strive for status in hierarchies and to engage in the kind of assertive, if not aggressive, risk-taking behavior that is often necessary to acquire resources and reach the top of hierarchies.

Male status is primarily a function of men's standing in the eyes of other men. One consequence of that fact is that men are inclined toward competition with one another even in circumstances in which it may seem objectively unnecessary or even unreasonable. Psychologists Martin Daly and Margo Wilson have described the "trivial altercations" that may occur between males, especially young ones, and their escalation into deadly violence:

> The emphasis on "triviality" obscures a still more important point. A seemingly minor affront . . . must be understood within a larger social context of reputations, face, relative social status, and enduring relationships. Men are known by their fellows as "the sort who can be pushed around" or "the sort who won't take any shit," as people whose word means action and people who are full of hot air, as guys whose girlfriends you can chat up with impunity or guys you don't want to mess with.[15]

Male status is thus often important not only for obtaining a mate, but for retaining her as well.

Some of the relationship between status and reproductive success may be obscured in complex modern societies. Comparison of reproductive success by socioeconomic class may lead us to discount the effect of status, since the higher classes may exhibit somewhat lower fertility than the lower classes, although their children may be more likely to survive to reproduce. It may be inappropriate, however, to view a modern society as a single breeding population for purposes of determining status. As Allan Mazur and his colleagues have observed, "The proper analogy to an animal dominance hierarchy is not socioeconomic status in mass society but the status hierarchy in a primary group of interacting humans who know one another."[16] It is generally within these smaller and more homogeneous groups that individuals compete for status and mates.

One confound of mating preferences is the relationship between status and resources. If a preference for either might be viewed as "primary," it would have to be a preference for status. For most of our evolutionary history, there has been no accumulated wealth; men varied in status—depending in large part on their skill as hunters, warriors, and leaders of men—but not in resources. In historic times, however, status and wealth have been highly correlated and often inextricably intertwined. To the extent that they are not entirely congruent, it may be status that is the more salient trait for female mate choice. A finding that financially secure women say they do not care about a prospective mate's financial resources would not mean that they do not care about his status. A well-respected yet financially struggling poet may be deemed a suitable match for a financially successful woman, though an unemployed lottery winner or an unemployed poet whose genius has gone unrecognized may not be.

Psychologists Alice Eagly and Wendy Wood have recently challenged the notion that women have an evolved preference for men having high status and substantial resources, arguing that the preference is simply a contingent strategy based upon the social fact that women lack economic power and therefore must secure economic resources through marriage.[17] They rely on the finding that, cross-culturally, the strength of women's preference for men with resources varies (albeit weakly) with the economic status of women in the culture. Their conclusion seems to be based upon the familiar, but misguided, view that if environmental influences can be shown, then biological influences are refuted. All they have shown, however, is that the strength of the preference is not invariant throughout the world irrespective of cultural conditions. But who would have expected it to be, even if the preference reflects an underlying biological predisposition?

There is no contradiction in saying that human mating strategies are products of evolved psychological mechanisms and also that they are somewhat environmentally contingent. Many animals far less able than humans to make the conscious utility calculations that Eagly and Wood ascribe to women modify their mating behavior in response to environmental factors. In the sex-role-reversed pipefish, for example, copulation is ordinarily preceded by a conspicuous ritualized dance, at the conclusion of which the female transfers her eggs to the male's brood pouch and the male then fertilizes the eggs.[18] Males generally prefer to mate with large females, but in the presence of a predator they become less choosy, no longer showing a preference for larger females. Similarly, although female field crickets generally prefer males with long calls, they will settle for a male with a shorter call if they would have to cross an open area to get to the long-calling male.[19] We would not say that the preference of male pipefish for large females and the preference of female crickets for long-calling males are cultural constructs simply because they vary with environmental conditions. Unless one believes that biologically influenced behaviors must be expressed as fixed action patterns irrespective of the environment, there is no reason to interpret Eagly and Wood's results as a refutation of the biological argument.

Eagly and Wood's argument implies that if the economic differences between men and women disappeared, so too would the differences in their mate preferences. Existing data do not support that prediction. Women who have greater economic resources or potential actually appear to place more importance on a man's economic status than do women with fewer resources,[20] and although the sex difference in importance attached to a potential partner's financial prospects has decreased in recent decades, that decrease is attributable largely to its increasing importance to men rather than its decreasing importance to women.[21] Moreover, marriages in which the wife out-earns the husband tend to be at especially high risk of dissolution.[22]

Female Reproductive Success: Nurturance versus Status

Unlike men, women cannot generally enhance their reproductive success by acquiring wealth or accumulating mates, and in some cases it appears that women undermine their reproductive success by acquiring political status.[23] As primatologist Barbara Smuts has suggested, competition among females is at a low level because "the outcome of a single interaction rarely leads to large variations in reproductive success because female reproductive performance depends mainly on the ability to sustain investment in offspring over long periods of time."[24] In contrast to men, then, women have increased their reproductive success by devoting the bulk of their energies to investment in children, through provision of milk and other forms of direct caretaking, rather than through acquisition of resources.

There is reason to believe that the closer connection of women to their infants is an evolved response, as a general tendency in a mammalian mother to be indifferent to separation from her infant would have been highly disadvantageous.[25] An intriguing line of evidence implying sex differences in the factors that activate parental feelings and behaviors are studies finding that among nongenetic parents, women's attachment may suffer more than men's. For example, stepfathers tend to have better relationships with their stepchildren than stepmothers do and are more likely to report parental feelings.[26] Similarly, the relationship of adoptive mothers to their adoptive children seems to resemble the genetic mother-child relationship less than the relationship of adoptive fathers resembles the genetic father-child relationship. A study by psychologist Irwin Silverman and colleagues examined perceptions of parental solicitude among adults who had been raised either by birth parents or adoptive parents.[27] Not surprisingly, birth children tended to perceive their mothers as the more solicitous parent. However, adoptees reported substantially less parental solicitude from their mothers than birth children did, but there was no such decline in solicitude between birth and adoptive fathers. Indeed, adoptive fathers tended to be rated higher in solicitude than adoptive mothers. These findings can plausibly be interpreted to mean that the experience of carrying and giving birth to a child predisposes women toward later nurturing feelings and behavior toward their children, while for fathers other factors trigger these feelings. Psychologist Geoffrey Miller has suggested that in our ancestral environment women commonly had children by successive males and that an evolved willingness to invest somewhat in stepchildren (although less than in their own biological offspring) may have been selected for as a mating tactic.[28]

Substantial reproductive tradeoffs for female competition and aggressiveness may limit the development of dominance in females. Among baboons, for example, high-ranking females obtain some clear reproductive benefits as a consequence of enhanced access to nutritional resources: they have higher infant survival, shorter interbirth intervals, and daughters who usually give birth at a younger age.[29] Their lifetime reproductive success, however, may not substantially exceed that of less-dominant females. Dominant females have greater miscarriage rates and may show signs of reduced fertility. Thus, the same causal factors that lead to high dominance may also carry reproductive costs that have acted as a constraint on selection for competitiveness among females. That common factor is likely to be testosterone.

The same effect may occur in humans. It is often reported that female executives have fewer children than male executives and fewer than the average woman. A study of law school graduates found that 40 percent of the women remained childless fifteen years after graduation.[30] The usual implication of these findings is that women must choose between work and family and that these women chose work. There is another possible explanation, however. Women who succeed in business tend to be relatively high in testosterone, which can result in lower female fertility, whether because of ovulatory irregularities or reduced interest in having children. Thus, rather than the high-powered career being responsible for the high rate of childlessness, it may be that high testosterone levels are responsible for both.

Parental Treatment

In chapter 7, we discussed how boys and girls seem to be predisposed to act in certain ways. Children are not alone, of course, in acting pursuant to their own distinctive psychologies. It is possible, and perhaps even likely, that when parents differentially reinforce behaviors in boys and girls, they too are acting pursuant to evolved psychological mechanisms. Human young have an exceptionally long period of dependence on their parents, and parents expend great effort in training their children. Mechanisms predisposing parents to inculcate sex roles in offspring may have evolved in tandem with children's predisposition to learn them. Parents do seem to reinforce some sex-typed behaviors, not just in our society but in all societies, although the extent of the reinforcement varies. Behavioral ecologist Bobbi Low has shown that the variation has a systematic pattern to it that is consistent with the predictions of evolutionary biology. She found that in polygynous societies, where the potential reproductive payoff of male competition is highest, parents train their boys to be especially competitive.[31]

Battle of the Sexes?

Sex differences in reproductive strategies can create substantial conflicts between individual men and women. As a rough-hewn generalization, men are much more often than women looking for unencumbered sex; women are more likely to be looking for an investing mate. The significantly greater costs to women from engaging in sexual intercourse is probably responsible for women's apparently greater inhibition of sexual arousal. A woman ready and willing to have uncommitted intercourse with the first man who flashed her would be unlikely to rear many offspring to reproductive age.[32] As Barbara Smuts has observed, "male interest in mate quantity, combined with female interest in mate quality, creates a widespread conflict of interest between the sexes."[33] This difference in goals often leads to conflicts between individual men and women, and, as we will see in chapter 13, in the workplace it can lead to the form of conflict known as sexual harassment.

Sex differences in reproductive strategies are often taken as reflection of a "battle of the sexes," and in a sense they are. The metaphor is misleading in important respects, however, because it implies that the interests of members of each sex are largely congruent with the interests of others of their sex. In fact, most competition, whether directly reproductive or otherwise, occurs between members of the same sex.

A conflict of interest exists between the male and female of potential or actual mating couples, but the male and female are not in any sense representatives of their sex. All males do not benefit if another male gets his way, and all females do not benefit if another female gets hers. Indeed, same-sex competitors generally benefit if the individual does *not* get his or her way. Moreover, the genetic interests of a male are often shared by his female relatives, and the interests of a female may be shared by her male relatives. Thus, cultural practices, such as purdah or infibulation, that are often perceived, especially by outsiders, as serving male interests to the detriment of female interests, often enjoy considerable female support.[34]

Intra-sexual competition often can have the superficial appearance of inter-sexual competition. For example, an elegant experiment by geneticist William Rice prevented female fruit flies from evolving by replenishing the supplies of females from the original stock while allowing males to evolve in subsequent generations. Rice found that when the females were prevented from evolving in response to changes in the male there was a substantial reduction of female survivorship, because the seminal fluid of the males became increasingly toxic as they developed chemical means of sperm competition.[35] This experiment is sometimes invoked as an example of competition between males and females, yet the males were not competing against the female but rather against each other. This inter-male competition produced harmful side effects to (that is, conflict with) the female, but the *competition* Rice observed was no more between males and females than the Battle of the Bulge was a competition between Germany, the United States, and Britain, on the one hand, and Belgium, on the other. Just as Belgium was merely the locus of the competition between Germany and the Allies, Rice's female fruit flies were the locus of competition among males.

Evolution and Cognitive Differences

An evolutionary explanation for the cognitive differences described in chapter 3 is quite plausible for some and less obvious for others. That men, the traditional hunters, should have greater dynamic spatial perception and targeting accuracy, for example, is hardly surprising.[36] But what of the difference in navigational technique? Irwin Silverman and Marion Eals have suggested that navigational differences can be attributed to the different demands of hunting and gathering.[37] Because hunters must pursue prey and then return home, there would be a substantial premium on a sense of direction that allowed them to proceed directly home after the kill, so that they would not have to retrace what might have been a very lengthy and circuitous route followed in search of prey. On the other hand, gatherers often return from year to year to the same location to find reliable sources of foodstuffs, a task that landmark recognition would substantially facilitate. Thus, while both hunting and gathering require navigational skills, they require somewhat different ones.

The Silverman and Eals explanation for sex differences in spatial ability in humans rests heavily on the sexual division of labor. Yet sex differences exist in many other mammalian species that lack a division of labor. Most adult mammals, especially noncarnivores, procure their own food. However, males often have larger home ranges than females, not because they wander in search of food but because they wander in search of mates. This pattern led anthropologist Steven Gaulin and biologist Randall FitzGerald to hypothesize that sex differences in spatial ability may not exist in strongly monogamous species.

In a clever test of their hypothesis, Gaulin and FitzGerald studied two closely related species of rodents, the polygynous meadow vole and the monogamous pine vole. They found that, indeed, the male meadow vole had a range overlapping the range of several females, while the ranges of male and female pine voles were largely coextensive. They then tested the voles on a radial-armed maze, which is a test of spatial ability. As predicted, a substantial sex difference in maze performance was found in the meadow vole but not in the pine vole.[38] This result does not definitively establish that sexual selection is responsible, because it is possible that these males have higher spatial ability precisely because they wander more. However, the hormonal evidence discussed above suggests that at least a substantial amount of the ability may predate the practice.

Psychologist David Geary has argued that male spatial superiority in humans has its origins in sexual selection, rather than in "ordinary" natural selection.[39] To the extent that male spatial prowess is an aid in warfare or in hunting—skills that are related to the number of wives that a man can obtain in many societies—it should be viewed as an outcome of "mating effort" rather than the "parental effort" that the traditional natural selection model would propose. Some intriguing support for this proposition comes from Kristen Hawkes's studies of hunting activities in hunter-gatherer societies.[40] One of the puzzles faced by behavioral ecologists is why men in many hunting societies devote so much effort to the bagging of large game, when hunting or trapping small game would provide a greater caloric return per hour. The hunter and his family often get no more, and sometimes less, of the meat than individuals not involved in the hunt. While in some circumstances, reciprocal food sharing can be seen as a method of reducing the variance of a chancy enterprise—I share with you today so you will share with me tomorrow—reciprocity in these transactions is often hard to find. Hawkes hypothesizes that men hunt large game for the status that bringing home a large mammal yields and that hunters share as a form of display because of the attention—including sexual attention—that the display attracts. Thus, Hawkes suggests, hunting and subsequent meat sharing may be viewed as a form of mating effort rather than parental effort. There is no reason in principle, of course, that successful hunting and sharing cannot serve both functions.[41]

Some have argued that the hunting/navigation hypothesis is illogical because women's foraging would also impose spatial demands.[42] One set of investigators has gone so far as to argue that for the hypothesis "to be valid, females must not have foraged away from the base camp."[43] That argument is equivalent to saying that men's superior throwing ability cannot be a consequence of selection for hunting skills if a woman ever threw a rock at an animal. The question is not whether any tasks performed by women require spatial skills—after all, women do have spatial skills—the question is whether selective pressures different in kind or in intensity were acting on the two sexes in the course of our evolutionary past.

An evolutionary explanation for sex differences in mathematical ability is less obvious than the explanation for spatial ability. While our ancestors have had to solve navigational problems for eons, for most of human history we have been blessed with total ignorance of differential equations. Mathematical ability, like reading ability, however, is presumably related to some other abilities that were important even before the development of mathematics or written language. David Geary has argued that mathematical ability is to a large extent a byproduct of spatial ability, based on studies finding that the male mathematical advantage disappears when spatial ability is controlled.[44] Others have expressed doubt because of the low correlation in other samples between spatial and mathematical ability.[45]

Philosopher Jeffrey Foss has adopted an extreme form of the anti-evolution argument. He argues that the suggestion that males have an innate superiority at mathematics necessarily must fail both because "infants have scant mathematical skills and those are equally shared according to gender" and because "mathematics is a recent socially imparted skill of this species, and it is not credible that evolution has already linked to the Y chromosome brain structures supporting mathematical skills."[46]

Both of Foss's arguments are fundamentally illogical. It is true that young infants of both sexes have virtually no mathematical skills, but they also have virtually no verbal skills. To say that males cannot in principle have an innate superiority in math means that females cannot in principle have an innate superiority in verbal skills. Similarly, there are no sex differences in infants' libido as far as we know, no sex differences in bodily hair, and scant sex differences in size and physical strength. By Foss's reasoning we must rule out the possibility that male adults, for biological reasons, are larger and stronger, have more body hair, or have stronger sex drives. Foss's second argument is no more logical. The innate human capacity to learn algebra has probably not evolved in the last millennium. There is every reason to believe that five thousand years ago humans were just as capable of learning algebra as modern humans provided the right educational background. Thus, there must have been some preexisting capacity to do algebra; if there were not, of course, then it could not have been invented. We can debate what that capacity was, and scholars do. But until we can say that the capacity was one that was not already differentiated by sex, the argument that there has been insufficient time for sex differences to evolve is simply misguided.

The adaptationist argument for women's greater verbal abilities is less developed. Some have suggested that the close relationship of mother and child has led to selective pressure on females for increasing verbal abilities. Others have argued that women's lesser strength put a premium on their ability to persuade verbally, what might be dubbed the *Scheherezade effect.* Neither of these arguments is yet convincing. The question is not, after all, whether verbal skills would have been important to women; rather, the question is whether they would have been sufficiently more important to women for selection to have conferred a verbal advantage on them. The primary reason for doubt is that verbal skills are also very important to men's reproductive success. They facilitate negotiation of male status hierarchies and assumption of positions of leadership. Male oratory is highly prized in many traditional societies, which would lead one to expect substantial reproductive payoffs to high levels of verbal ability in men. Another possible explanation for the sex difference in verbal ability is that there is a tradeoff between high spatial ability and high verbal ability, so men's lesser verbal ability may simply be a byproduct of selection for spatial ability.[47]

Those who acknowledge the greater physical strength of men and its biological origin but deny the existence of evolved temperamental differences are in something of an awkward position. If greater male strength is an adaptation, it must be an adaptation for something. If it is an adaptation for male-male competition, as it is in most species, it would be surprising if that physical dimorphism were not also accompanied—again, as it is in most sexually dimorphic species—by temperamental and behavioral concomitants. As Lionel Tiger has pointed out, the brain evolved not to think but to act.[48] If male physiques are adaptations to facilitate physically aggressive and dominance-seeking actions, that implies something important about the male psyche as well.

Critics of evolutionary explanations for particular traits often deride those explanations as "Just So Stories," like "How the Camel Got Its Hump" or "How the Leopard Got Its Spots"—a parlor game that anyone can play. Sometimes the criticism is valid, but often it is not. Speculation about the adaptive origins of traits often generates hypotheses that can be subjected to empirical testing. However, the critics who so cheerfully ridicule biological explanations do not seem to recognize that the social constructionists are no less likely—indeed, they are far more likely—to provide speculative explanations. Both evolutionists and social constructionists are interested in origins; evolutionists often attempt to provide a plausible and specific explanation for how a particular trait might have been adaptive, thus leading to testable hypotheses. Observing a difference and asserting on no evidence that it is socially caused lacks even the explanatory power of a "Just So Story." It is no more creative than simply asserting that a trait is biological without providing any explanation of its proximate or ultimate origins.

Evolved differences between the sexes have numerous social effects. As we have seen, one major locus for the operation of these differences is the workplace. Given the manifest sex differences in workplace outcomes, the question becomes what, if anything, can or should be done in the policy realm to respond to these effects. That is the subject of part IV.

IV Public Policy and Sex Differences in Workplace Outcomes

Having examined the reasons for the glass ceiling, the gender gap, and occupational segregation, as well as evolved sex differences that contribute to these phenomena, we are now in a better position to consider whether policy intervention is appropriate, and, if so, what form it should take. The threshold inquiry, in chapter 10, is whether these phenomena are actually the indicators of female disadvantage in the workplace that they are often taken to be. It is certainly arguable that they are not—that women make choices based on their priorities and are no less likely than men to choose a course that is optimal for them. "Optimal" in this context does not mean the best outcome in the best of all possible worlds; it means the best outcome in *this* world—a world of necessary constraints, sometimes requiring difficult tradeoffs.

Critics of current workplace arrangements offer solutions falling into one of two somewhat overlapping categories: those intended to mitigate the effects of perceived discrimination by employers or simply to achieve statistical parity for its own sake and those intended to mitigate the effects of the work/family conflict that disproportionately affects women's work lives. Proposals in the first category, which are the subject of chapter 11, include affirmative action to increase the number of women in high places, programs to increase the number of women in "nontraditional" jobs, and equal pay for "comparable" jobs to increase women's compensation. Proposals in the second category, the topic of chapter 12, include policies relating to part-time work, maximum hours, day care, and parental leave, as well as calls for husbands to increase their share of housework and child care in order to free more time for women's labor-market activity.

Many of the proposed policies would have little effect on our primary measures of workplace equality, and, indeed, some would have negative effects, although they may be desirable for other reasons. Others risk imperiling the relatively free market in labor that has made the United States a world leader in advancing women in the workplace.

10 Difference or Disadvantage?

Sex differences in workplace outcomes are typically viewed as indicators of a female disadvantage crying out for a remedy. The numerous tradeoffs that individuals make, however, such as those between work and family, mean that measuring disadvantage along a single dimension, such as earnings, provides no insight into any global measure of advantage or disadvantage.

The equal representation of the sexes among those who are career primary and those who concentrate their efforts at home is desired by some critics, but the desirability of that outcome should turn on how well it accords with people's preferences, and it does not appear that either men or women generally endorse that egalitarian goal. Indeed, women tend to be more satisfied with their jobs than men, and their greatest dissatisfaction is with the number of hours they work, even though they work substantially fewer hours than men.

It is commonly said that women lack choices, because women who wish to accommodate work and family are "forced" into marginal work. The existence of tradeoffs and constraints does not refute the existence of choice. If anything, women have a broader range of socially sanctioned workplace/family arrangements than men have, and men face the same tradeoffs that women do. The consequences of these tradeoffs differ between the sexes, however, such that the psychic costs of being separated from their children are higher for women than men, and the psychic costs of diminished workplace involvement and opportunity for advancement are higher for men than for women. It is therefore unsurprising that they tend to choose differently.

The tendency of women, rather than men, to be the partner to stay home if one does is often blamed on the unwillingness of men to adopt that role. It is probably less an issue of supply than it is of demand, however, as women with high-powered careers tend not to be attracted to the men who would fill that role.

Are the glass ceiling, the gender gap, and occupational segregation simply facts of the workplace, or are they problems that require some correction? Most workplace literature assumes that they are problems, often without being very explicit about why. Discrimination is invoked as an explanation so frequently that it is not clear whether it is the outcomes that are viewed as unacceptable or the processes presumed to be responsible. Thus, we must decide with some specificity what the problem is, and only if we identify one, is corrective action appropriate.

Are Women Disadvantaged in the Modern Workplace?

The first question is whether the differences that we see are merely differences or whether they are "disadvantages." Much of the literature on the status of women in the workplace assumes that statistical disparities are by definition indicators of both disadvantage and the need for change.[1] As we have seen, however, that premise is often derived from a too-narrow focus on the economic aspects of jobs to the exclusion of other factors either inside or outside the workplace. Still, it cannot be denied that women are less likely to achieve the highest levels of employment hierarchies—although some do—and that the average woman earns less than the average man—although millions of women out-earn millions of men—and that women are "underrepresented" in some occupations, though "overrepresented" in others.

How Do We Measure Disadvantage?

A persuasive case that difference amounts to disadvantage requires evidence rather than assertion, but what kind of evidence should we look for? Once we acknowledge tradeoffs among various job features, as well as tradeoffs between work and family, how do we determine if one set of individuals is disadvantaged relative to another on the entire constellation of job features? The problem is that there is no common metric by which to assess the various tradeoffs that workers make. Assume that three women holding $50,000-per-year jobs all get pregnant at the same time. Woman A takes a two-month maternity leave and then returns to work full-time; Woman B takes a twelve-month maternity leave and then returns to work part-time; Woman C leaves the work force entirely. Five years later, Woman A is earning $100,000 per year but isn't sure what her child is doing in kindergarten; Woman B is earning $25,000 per year and is home every day when her child comes home; Woman C has no earnings, but is the room mother for her child's kindergarten class. These are not identical outcomes, but are they "equal" or "unequal"? If they are "unequal," how can we rank order them? To say that Woman A's outcome is $75,000 better than Woman B's, whose outcome is in turn $25,000 better than Woman C's, is such a unidimensional comparison that it is meaningless. The best that we can say is that each woman made a decision based upon her own situation and her own preferences, and we must assume that each one obtained the optimal (for her) result. Similarly, if a mathematically talented woman chooses another field because her social orientation makes the cloistered life of a professional mathematician or scientist unappealing, should we view her as a victim of "cold" science or an active participant in the construction of her life?

If we look at just men or just women, we see that within each sex there are examples of just about all conceivable roles. Both men and women occupy the roles of stay-at-home spouse, part-time worker, full-time worker who has cut back for family reasons, and full-throttle full-time career-committed worker. Is it morally problematic that different individuals make different choices? That they do seems the essence of choice. Is it morally problematic that the stay-at-home husband receives no weekly paycheck, that the part-time male worker receives a smaller check than the full-time male worker, or that the full-time male worker has less time to spend with his children than the stay-at-home father? These outcomes seem "natural" products of our economic system, although perhaps that is just because we have gotten used to them. Is the stay-at-home husband oppressed relative to men in the other categories? The fact that he stays home deprives him of the economic and noneconomic benefits of labor-market participation, yet it also provides him with compensating benefits.

If people do not view these individual differences as morally problematic, then what accounts for the offense that so many take to the fact that as the continuum progresses from stay-at-home spouse to full-time career-committed worker, the proportion of men increases and the proportion of women decreases? Would the objections disappear if the proportion of men and women were equal in each category? It appears that many of them would, suggesting that despite rhetoric to the contrary, the fact that individual workers must make tradeoffs between career and family is not by itself perceived as a problem.

Some writers have explicitly extolled a world in which men and women are equally likely to fill all roles. Rhona Mahony argues, for example, for "throwing away the stereotypes" that lead men to do less than half the work of raising children.[2] The key to workplace equality, she argues, is abolition of the sexual division of labor at home:

> When the sexual division of labor has disappeared, men will be doing half the total amount of child-raising work. Roughly half the primary breadwinners will be women and roughly half will be men. Roughly half the homemakers will be women and roughly half will be men. Also, it will mean that roughly half the primary parents—the ones who stay home when Junior is sick, who carpool to soccer practice, who cook chili for the bake sale—will be men. Those men will be economically dependent on their wives. They'll do what millions of women have done for so long: they'll focus on their children.[3]

How Mahony expects society to exact compliance with its modified expectations is not clear. Society expects all sorts of things from its members—not driving drunk, not taking illegal drugs, not paying less than our allotted share of taxes, and so on seemingly endlessly. But the expectation is not enough by itself to guarantee compliance, so society must construct mechanisms to enforce its expectations and impose penalties on those who fail to comply. A free society is limited, however, in its ability to enforce its expectations about who stays home with a sick child or drives the carpool to soccer practice.

Why is a fifty-fifty distribution in each role morally superior to the sexually differentiated distributions that prevail today? The only reason that comes to mind is that an equal distribution may accord better with individual preferences than the current distribution does. Under this view, individual preferences are currently thwarted by social constraints, and people must be freed from these constraints to allow their preferences to operate.

It is in dealing with the issue of preferences, however, that the argument of many feminists often founders, because they have a hard time demonstrating that women actually want what some activists think they should want. If current arrangements are unfair and oppressive to women, one might expect that women would be dissatisfied with their work lives. In fact, however, job-satisfaction surveys consistently find that women are as satisfied as men, and often more so, implying that the workplace/family accommodations they have reached may be satisfying to them.[4] This equal or greater satisfaction applies even specifically to compensation despite the fact that women earn less than men.[5] The dimension along which women tend to be least satisfied is the number of hours they work, despite the fact that they work substantially fewer hours than men. On its face, then, the pattern of female satisfaction seems somewhat inconsistent with the notion of female disadvantage.

One might have thought that high levels of satisfaction among women would have been viewed positively by those who take it upon themselves to advocate for women. Instead, however, that satisfaction is a source of immense frustration. Law professor Deborah Rhode has declared women's happiness to be a problem that she has dubbed the "no-problem problem," and she complains that "a central problem in generating perceptions of a problem among women is that their opportunities have generally been expanding."[6] This characterization makes sense, however, only if the goal is the perception of injustice rather than its elimination.

What Do Feminists Think That Women Want?

The case for intervention is strongest if it is believed that there are systematic structural barriers to the satisfaction of women's preferences that do not exist for men. So, what is it that feminists think that women want? Do they actually believe that men's and women's preferences are identical? It seems not.

Many feminists acknowledge that women do not receive equal extradomestic outcomes in part because they have opted for a larger measure of satisfaction in the domestic sphere. Law professor Nancy Dowd has acknowledged, for example, that "for each sex, the ideal relation between work and family is constructed differently." She notes that the "clash between occupational and family life cycles has produced starkly different patterns of labor force attachment for men and women" and that women have much more frequent job interruptions and higher absenteeism rates than men.[7] "Women," she says, "continue to fit work to families, and men vice versa." Deborah Rhode observes that women's lesser commitment to the workplace is caused not only by their greater domestic role, but also by the fact that they "have placed lower priority than men on objective forms of recognition in employment such as money, status, or power."[8]

Well, now we have some common ground. These writers seem to be agreeing with the central point of this book: the two sexes have different priorities and make different tradeoffs. So, what separates their arguments from those set forth in this book? Primarily two things. First, they often deny the possibility that the choices women make are freely chosen; instead, they are "choices." Second, they do not believe that these "choices" should have economic consequences.

Do Women Have a Choice or Just a "Choice"?

If women are less willing than men to sacrifice family for career and therefore elect not to do so, their decisions might be considered free choices raising no serious public-policy question. Yet many writers refuse to acknowledge, let alone value, the fact that a woman might freely make the apparently selfless (though, as we know, genetically self-interested) choice to nurture her children. Instead, they deny that there is any true choice at all; at best there is "choice," the quotation marks indicating that any choice that exists is a forced one. Thus, Nancy Dowd asserts that for women "the choices are complete separation of work and family, or 'choosing' marginal work."[9] According to Joan Williams, scaling back work commitments to accommodate children's needs, remaining childless, and dropping out of the paid labor force are all "choices," rather than choices. These actions are clearly not forced in the legal sense of duress, so what is it about them that deprives them of the dignity of true choices? There are two reasons, says Williams: first, the range of choices is substantially constrained; second, it is usually the wife, rather than the husband, who scales back on work commitment.[10] While both of these contentions are true, neither demonstrates that the decisions are choices in only the ironic sense.

The world is full of constraints, but those constraints do not deprive us of free will. They do, however, make our lives more difficult. Time is finite, so time-allocation decisions are ordinarily zero-sum: an hour spent on activity A leaves one hour less to devote to activity B. Decisions are often difficult because each course has advantages and disadvantages and the costs and rewards of each decision are often in different currency. The decision whether to compromise career to spend more time with children is often a difficult one; either way, one gets something of value and gives up something of value. But to say that one takes the bad with the good is not to say that one is forced to take the bad. Christine Littleton apparently disagrees. She argues, for example, that if "women do in fact 'choose' to become nurses rather than real estate appraisers, . . . they certainly do not choose to be paid less."[11] Well, yes, they do, just as employees who move from full-time status to part-time status to be with their children choose to be paid less, even if they would prefer to suffer no wage penalty for decreasing their hours.

The fact that it is usually the wife rather than the husband also does not demonstrate that there is no true choice. In fact, that is precisely the outcome that one would expect from a regime of free choice if men and women have the different sets of preferences that have been described throughout this book. It should also be recognized that the complaint that it is usually the wife who cuts back to spend time with children implicitly contains a value judgment that not everyone would endorse—that the one who makes that choice gets the short end of the stick.

Why Don't Men Face the Same Choices as Women?

One source of resentment for some women is the perception that they are forced to make tradeoffs between family and career that men do not face. Williams, for example, asserts that "women must choose between work and family while men can have both."[12] When a wife leaves the work force, the husband is then "the only one who could 'have it all.'"[13] One might question exactly what the "all" is that he has: complete financial responsibility for his family and less time to spend with his children.

Although the tradeoffs are the same for men and women, the psychic costs and benefits of different choices are not the same because of differences in male and female psychologies. Men, on average, get greater satisfaction out of achieving power and status than women do. Women, after spending some time on the fast track, are more likely than men to say, "If this is what I have to do to succeed in this career, it simply is not worth it to me." Thus, the payoff to women for single-minded career commitment is smaller. Conversely, the psychic costs are higher, on average, for women than for men, in large part because women more than men dislike being separated from their children for most of the day. British writer Rosalind Arden has written of how, as a sometimes working mother, she "aches under the onus of having to choose between work and mothering" and of the pain experienced by her women friends in high-powered jobs who must significantly reduce the time spent with their children.[14] Where does that psychic pain come from, if not from the evolutionary legacy that has endowed women, at least when conditions are favorable, with a propensity to remain close to their children? Mammalian mothers have evolved not to be indifferent to separation from their infants. Men, especially ambitious ones, are significantly less likely to feel that constant tug between children and career, not because they love their children less but because they can compartmentalize in a way that women, on average, cannot, and they are prone to showing their love in a different way.

The assertion that women cannot have a high-powered career and a family but that men can is based upon a sexual asymmetry in what is meant by "having a family." For many women (and far more women than men), it means spending substantial time nurturing their children on a day-to-day basis, not simply ensuring that someone else does it. It is likely to mean being involved in school activities, seeing their children off to school in the morning, and being there when they get home from school. This difference is reflected in its extreme form by the linguistic distinction between "fathering a child" and "mothering a child." Only by failing to realize that the phrase "having a family" is different for the two sexes is it possible to argue that women are asking only for what men already have.

The unstated double standard only makes it appear that men are in a better position than women. A male executive who has a wife who does not work outside the home and who seldom sees his children is considered to "have it all"—career and family. Even if his wife works and his children are in day care all day, he still is viewed as having it all. The female executive—perhaps the wife of the male executive described above—whose children spend their days in day care and whom she seldom sees, is seen as having sacrificed family for career. She does not "have it all," even though she has exactly what her husband has.

Men and women face the same choice; the female executive can do the same as a man if she chooses. By taking only a brief pregnancy leave and placing the child-care responsibilities on someone else (for example, a paid provider or a willing husband), there is little reason that all else being equal the woman cannot have career and family the same way that a man can. But that is not good enough for many women because that is not what they want.

Are Women's Choices Really More Constrained than Men's?

There is some irony in the feminists' seizing on the lack-of-choice issue, since women have a much broader range of socially acceptable work/family choices than men do. A woman can choose to be a "career primary" worker, a "career and family" full-time worker, a part-time worker, or a full-time homemaker; all of these are socially respected choices. Men, on the other hand, have little choice at all. They are expected to be full-time workers who, in most circumstances, are the primary family breadwinners. This greater freedom of females to adopt male roles than vice versa has its origins early in childhood[15] and makes puzzling the argument that women are the primary victims of sex roles. If they are victims at all, it seems, they are victims of their own preferences.

Rather than identifying women as the victims of current arrangements, one could as easily argue that it is men who are disadvantaged.[16] Men are expected to work, whether or not their wives "choose" to. Their biologically predisposed competitiveness and status seeking can be as much a straitjacket as women's biologically conferred role as incubators of children, and they often lead men to an early grave. This is not to suggest that the mantle of victimhood should be lifted from women and conferred on men. What needs to be questioned is the notion that either sex is a victim. Once one accepts that men and women have, on average, different predispositions, desires, and definitions of success, it is axiomatic that comparisons between the two groups will yield qualitatively different outcomes for the two groups, at least in a society that permits people freely to pursue their own desires rather than requiring them to satisfy someone else's.

Why Does It Just "Happen" That It Is the Wife Who Stays Home?

Why is it usually the woman who chooses to leave the work force or cut back on work hours? To be sure, even if male and female preferences were identical, in today's culture the woman would probably be more likely to do so. That arrangement requires less explanation and is less likely to be met by the eyebrow raising that may attach to a man's being a "house husband." Moreover, because the husband is likely to have the larger income in the couple—if only because he is likely a little older and more established in his career—it may make economic sense for the couple to view the husband's career as primary. But these are only partial explanations. The fact is that staying home with the children is less likely to suit the temperament of men than of women.

This is not to suggest that men do not feel a strong attraction to their children; they clearly do. Accounts are plentiful of men who describe looking upon their newborn infants with the same kind of intense love that women often describe (and probably not coincidentally experience drops in testosterone levels).[17] But men, on average, differ from women in the extent to which this is an all-consuming feeling. Accounts of men who cannot concentrate at work, thinking about their baby, hoping it is OK, and worrying whether the child-care provider is doing an adequate job are far less plentiful than equivalent accounts of women. This "compartmentalization" means that men are less likely to feel the constant "tug" that so many women feel and that leads many women to dissatisfaction with both their work and home lives.

Why Aren't House Husbands as Available as Housewives?

Many men with high-powered careers have an affirmative preference for a stay-at-home wife, and there is no apparent shortage of women willing to fill that role. Are there many high-powered career women with a desire for a stay-at-home husband? It seems doubtful, despite the reported complaints of some that they "need a wife." Women, unlike men, tend to prefer mates who are at least equal, if not higher, in status to themselves. Because men's status comes primarily from their employment, women's desire for status in a mate is in substantial conflict with a desire for a husband who stays home. Thus, it seems unlikely that there are many female executives and professionals who would be eager to marry a man who would stay home with the children. John Townsend's study of medical students confirmed this view, finding that women expressed a strong aversion to having a husband who did not work outside the home, while men viewed having a nonworking spouse much more favorably.[18]

Well, one might argue, this is all moot anyway, because men are simply not willing to stay home for an extended period being supported by their wives and caring for their children. That is only partially true. Women with six- and seven-figure incomes could easily find men willing to be supported. These men, however, would almost by definition lack traits that women find highly desirable in a mate: dominance, ambition, and status. Again, among John Townsend's medical students, "no woman preferred a husband with lower status and income than she had, and no man preferred a wife with higher status and income than he had."[19] Thus, the problem for women is not that there are no men they can marry and get the equivalent of a housewife; it is that they really do not want one. Women's biologically ingrained mate preferences tend for the most part to preclude the house-husband model.

The different average preferences of men and women make it difficult to make the case for female disadvantage generally or, more specifically, the case that differential outcomes are necessarily—or even probably—a consequence of discrimination. The policy initiatives discussed in the next chapter, which tend to be based upon assumptions of disadvantage or discrimination, should therefore be given a skeptical reception, although they may be supportable under different rationales.

11 A Thumb on the Scales
Changing the Rules to Improve the Numbers

A number of policy initiatives have been adopted or proposed to attempt directly to equalize various workplace outcomes between the sexes—affirmative action, programs to integrate traditionally male occupations, and "comparable worth."

Affirmative action (meaning preferential selection by sex) is a popular countermeasure for all of the indicia of perceived inequality, because it can, by fiat, produce results different from those that would be obtained through normal processes. Even if affirmative action is legally permissible, there are reasons for caution. Individuals hired through affirmative action tend to be judged less competent than those selected on the basis of merit. Moreover, women selected preferentially tend to exhibit less self-confidence and a lesser willingness to take on difficult tasks than either men or merit-selected women. Because advancement in organizations often depends on a willingness to undertake challenging tasks, preferential selection may rob women of the confidence they need to advance further.

Efforts to attract more women to nontraditional occupations sometimes take the form of preferential selection, but other initiatives are aimed at increasing the supply of women qualified for and interested in the positions, such as encouraging girls to take more high-school science courses. Some universities are attempting to make science education more "female friendly" by providing special tutoring and other resources to female students, an initiative that may be inconsistent with antidiscrimination laws. Some argue that science needs to be "fixed," although one might question the extent to which a mathematical field such as physics can be modified to better suit the temperament and interests of people who are not mathematically gifted or who desire a high social dimension in their occupation. Efforts are also directed toward increasing the representation of women in such persistently male fields as firefighting. Because the interest of women in such positions is low and the physical demands are high, the effect to date has been modest.

Because little of the gender gap in compensation results from pay disparities among workers performing the same job, comparable-worth systems have been advocated to require employers to provide equal pay for very different jobs that re-

quire "comparable skill, effort, and responsibility." The primary objection to comparable-worth systems is that they divorce the wage rate from principles of supply and demand. Raising the wages of jobs for which the supply of workers is abundant and lowering the wages of jobs for which the supply of workers is limited creates unfortunate dislocations in the labor market. Even if the practical problems could be overcome, comparable worth would have only limited effect on the gender gap, because men and women tend to work not only in different jobs but also for different employers.

Many employment practices or policy initiatives are aimed at achieving a work force in which women are represented and compensated in a statistically equal way. These include affirmative action to increase the number of women in high-ranking and nontraditional positions, efforts to attract women to scientific and blue-collar occupations, and comparable-worth requirements to attempt to equalize male and female compensation.

It is often not clear whether advocates are concerned with the process by which the results are presumed to have occurred or simply the results themselves. The rhetoric behind most of the initiatives described in this chapter often includes a large measure of talk about discrimination, although some of the proposals have substantial support that may be independent of an anti-discrimination justification.

Affirmative Action

A frequently suggested nostrum for the various statistical disparities in workplace outcomes referred to throughout this book is preferential treatment by sex or "affirmative action." Affirmative action is viewed by some as an appropriate and effective means of increasing the number of female executives and professionals, increasing the number of women in nontraditional occupations, and reducing the gender gap in compensation.

Many debates over the propriety of affirmative action get bogged down in semantic distinctions, often resulting in partisans' talking past each other. Terms such as *affirmative action, quotas, outreach, preferences, goals and timetables, reverse discrimination, plus factors,* and *tiebreakers* are given varying definitions, allowing disputants to make arguments that are internally consistent but that, because of shifting definitions, fail to engage the arguments of their opponents.[1] In its most benign and noncontroversial senses, affirmative action means ensuring that the employer is not discriminating and that it is "casting a wider net," broadening the pool of potential applicants but treating those who apply in a sex-blind manner. In this section, however, the term is used to mean the more controversial and arguably less benign (but nonetheless widespread) practice of making sex-conscious selection decisions that deviate from the employer's ordinary selection criteria.

The Justification for Affirmative-Action Plans

Three primary justifications for affirmative action are put forward by proponents. The first is the "role model" theory, which holds that in the absence of female role models in certain jobs, girls will not aspire to these positions and they will remain predominantly male. There is little evidence, however, that the existence of such role models will actually alter girls' preferences. The second is "diversity," under which it is asserted that inclusion of women will guarantee a wider variety of perspectives and permit the organization to function more effectively. There is also little evidence that in most organizations this kind of diversity has the effects attributed to it. The third justification is a remedial one; affirmative action is said to be necessary to remedy historic discrimination against women, although as we have already seen many of the statistical disparities sought to be redressed have little to do with discrimination.

Whether or not the justifications for the role-model or diversity rationales can be supported empirically, neither of these rationales is a legitimate basis for preferences under current Supreme Court doctrine. Instead, sex preferences are permissible only to undo the effects of discrimination. Although, at least in the private sector, the discrimination may be a society-wide discrimination, rather than merely discrimination by the particular employer, the Supreme Court has emphasized that sex, if used at all, must be used "in a manner consistent with Title VII's purpose of eliminating the effects of employment discrimination."[2] Nonremedial use of sex is limited to those narrow circumstances in which sex is a *bona fide occupational qualification,* where the employer is, in essence, compelled because of the nature of the job to hire employees of one sex.[3]

The predicate for remedial affirmative action for women throughout the economy is weak, even if it may be powerful in some limited contexts. Even assuming, however, that an adequate justification for sex-based preferences exist, there is a further question whether they would achieve their goals. In fact, there is good reason to believe that they impede some of the very goals they are intended to further.

The Effect of Affirmative Action

Opponents of affirmative action have long argued that it is harmful to beneficiaries and nonbeneficiaries alike. Beneficiaries are harmed by the stigma of incompetence that attaches to a person believed to have been selected on criteria unrelated to competence.[4] Members of the preferred group who obtained their positions purely on the basis of merit are presumed to have received their jobs through preferences, so the taint of affirmative action applies, albeit unfairly, to them. Members of nonpreferred groups who lose out in the selection process tend to feel that they were treated unfairly, often leading to resentment that can interfere with acceptance of members of the benefited group. Moreover, those who know they were selected preferentially may question their own competence. A fairly large empirical literature supports the validity of these concerns.

The Perception of Others about the Competence of Beneficiaries. A substantial body of research, much of it produced by psychologist Madeline Heilman and her colleagues, validates concerns about stigmatization of affirmative-action beneficiaries. One study found that when told that a woman is an affirmative-action hire, both males and females tend to rate her as less competent than either a woman chosen on the basis of merit or a man.[5] This effect is found for both jobs that are strongly male-sex-typed, where women's competence would already be suspect, and those that are only weakly so, where women would otherwise not be viewed skeptically. In either case, the importance of the individual's qualifications in the selection process seems to be discounted by knowledge that sex was important to the decision.

Because individuating information typically diminishes reliance on stereotypic beliefs, one might expect that subsequent competent performance would counteract the initially low estimation of affirmative-action hires. Indeed, in some circumstances it does. When performance information is clear and unequivocal and where good performance is clearly attributable to the individual, then the method of selection recedes in importance.[6] In contrast, if indicators of success are present but ambiguous, preferentially selected women are rated no higher in competence than they are if either no information or even information conveying failure is provided.

Many jobs—especially high-level ones—do not provide the unequivocal information that seems to be necessary to dispel the taint of preferences, because assessment of competence is often subjective. Moreover, successful performance in many jobs may require the contributions of many people. Unless success is clearly attributable to the individual in question, rather than to the quality of the people under her, the taint apparently persists because people will tend to think that she is simply lucky to have such competent people working for her.

The Effect on Self-Confidence of Beneficiaries. Perhaps the most serious effect, and the one that may be the least intuitively obvious, is the effect of preferences on those who receive them. In a series of laboratory studies, Heilman and her colleagues have shown substantial adverse effects of sex-conscious selection.[7] Because preferential selection seems not to provide the reinforcement of competence that merit selection does, those who require reassurance of competence in order to maintain confidence are disadvantaged by knowledge that they were selected preferentially.[8] As we saw in chapter 2, in the absence of information about ability, women tend to assume low ability and men assume high ability. Moreover, women who know that others view them as having been selected preferentially infer that others have lowered expectations of their competence.[9]

The affirmative-action label influences both self-perceptions and behavior. In a series of laboratory studies, Heilman and her colleagues have found that almost all men have confidence in their abilities and are willing to select difficult tasks whether they believe they were chosen on the basis of merit or on the basis of sex. Although women selected on the basis of merit respond very much like the men, women selected on the basis of sex demonstrate substantial diminution of confidence and a much greater inclination to select easier tasks.[10]

The detrimental effects of preferential selection can be mitigated somewhat by provision of unambiguous merit information. Heilman and colleagues told five groups of female subjects that they were chosen on the basis of merit, preferentially selected over equally qualified men, preferentially selected because they satisfied some minimum standard, preferentially selected with the criteria for selection left ambiguous, or automatically selected without regard to qualifications because they were female.[11] Except in the circumstance where sex was used as a tiebreaker, preferential selection had the same depressing effect on women's evaluation of their performance and willingness to persist in a leadership role regardless of the extent of the preference. However, women who were told that sex was used only as a tiebreaker evaluated their performance and leadership ability the same as merit-selected women, although their desire to relinquish the leadership role was as strong as that of other preferentially selected women. The perception of others that preferentially selected women were less competent was diminished but not eliminated by knowledge that the preference extended only to breaking a tie. The researchers also found that when the nature of the preference was left ambiguous, subjects reacted as if merit criteria had been totally disregarded.

This study suggests that using sex only as a tiebreaker can avoid some, though not all, of the negative consequences of preferential selection. However, unambiguous information that a person preferentially selected is as qualified as the competition, not merely adequately qualified, is relatively rare in real life. In most settings in which sex preferences are employed, they operate as more than simple tiebreakers. Instead, from among a group of candidates who are deemed qualified, the employer gives weight to sex, so that merit selection would produce one result and preferential selection another. If a woman perceives this fact, then she is likely to be negatively affected by the preference, as were Heilman's subjects who were told they were minimally qualified or selected through an ambiguous process.

The implications of these and similar studies are potentially profound. They suggest that women given preferential treatment are likely to have less confidence in their abilities and to shun challenging tasks. Advancement in organizations depends not just upon how one performs the tasks that one undertakes, but also on what kind of tasks are undertaken in the first place. Seeking challenging tasks is a sign of commitment, and good performance on such tasks is a better indicator of competence than good performance on less-demanding tasks.[12] Thus, preferential advancement may actually sap women of the confidence that they need to do well and advance even further.

The question, then, is what employers can do to mitigate the problem of the lowered perceptions of ability that attend preferential selection. The employer can strongly emphasize that selections were made on the basis of merit, but such assurances—particularly if made only when hiring women—may just create suspicion that the employer "doth protest too much." Employers could, of course, simply pretend that they do not have affirmative-action programs, that sex plays no role in their employment decisions, and that all decisions are based purely on qualifications. In many circumstances, however, such as where selection criteria are relatively objective and well known, an employer could not plausibly make such a claim. Even if it could, there are many reasons not to do so. The employer stands to lose credibility if its assertions are revealed to be untrue, but even beyond that is the risk of legal liability. Employers that engage in sex preferences pursuant to a bona fide affirmative-action plan may be entitled to a measure of immunity against legal liability. Ad hoc reliance on sex, not on the basis of an affirmative-action plan, however, is not cloaked with that immunity. Thus, even if the employer were inclined to be dishonest about what it was doing, it would be a legally risky strategy. Moreover, such coyness about the existence of affirmative action would deprive employers of much of the public-relations benefit of aggressive affirmative-action programs. Many companies tout their affirmative-action plans not only within the company but also in the community. Indeed, public statements by executives that diversity and affirmative action are not just the right thing to do but also "good business" have become de rigueur in the corporate world.

The case for affirmative action has yet to be made. Political and legal support is heavily dependent upon its role as a remedy for discrimination, but, as we have seen, many of the statistical disparities toward which it is directed have little if anything to do with discrimination. Regardless of the cause of the disparities sought to be reduced, however, affirmative action may create more problems for its intended beneficiaries than it solves.

Attracting More Women to Nontraditional Occupations

Many people advocate measures to increase the number of women in nontraditional occupations. Often left unstated is the threshold argument for why we should care about the sex ratio in these occupations assuming that the selection process is a fair one. That is, if women tend not to be interested in certain jobs, it is not clear why that fact should be any greater reason for policy intervention than the fact that men are not interested in others. To be sure, if arbitrary barriers or pervasive misperceptions are excluding women from jobs, it is sensible to eliminate those barriers or misperceptions, but the mere fact of lack of interest is not self-evidently a problem.

Increasing the Number of Women in Science

Women are, as we saw in chapter 5, well represented in some scientific disciplines but still relatively rare in others. The larger the social dimension of the discipline and the less it imposes high-level mathematical demands (not wholly independent factors, of course), the more women tend to be found in the discipline. The observed patterns seem far more reflective of self-sorting than they do of some general hostility toward women in science.

Some calls for increasing scientific training for women are based upon the relatively poor showing that the United States makes in international comparisons of scientific literacy. Increasing the participation of women in science, it is said, would expand the talent pool. Certainly it is in the national interest to increase scientific literacy generally and to strive for scientific excellence. It is not clear, however, that a general increase in scientific rigor in the curriculum would diminish the sex disparity. If anything, as we saw with the TIMSS results in chapter 7, increasing the rigor of math and science education would probably *increase* the sexual disparity.

Even the most ardent egalitarians have not yet explicitly advocated lowering the level of scientific literacy in the country to reduce the sexual disparity. Some of their policy proposals, however, may have much the same effect. The National Council for Research on Women, for example, has recommended elimination of the difficult "gateway" courses in computing, physics, and engineering and replacing them with courses that "invite" students into the disciplines.[13] Such a course of action creates a risk of curriculum dilution, not to mention a risk of misleading students about the true nature of the discipline and causing them to expend a greater amount of time learning that the discipline may not be for them.

If certain scientific fields are somehow a less-perfect fit for women than they are for men, then three approaches suggest themselves—changing women, changing schools, or changing science itself (or some combination thereof).

Changing Women. One commonly urged intervention is to get girls to take more math and science courses in high school to provide them the underlying foundation for more advanced study. As we have seen, however, up until calculus, more girls than boys take math courses, and even in calculus, the difference is trivial. Moreover, while more boys than girls take physics—a course generally recommended for college engineering study—more girls than boys take high-school chemistry, also a recommended course for college-level engineering study. A study based on a nationwide sample found that if girls took as many high-school math and science courses as boys, the proportion of women who would choose engineering would increase by just 0.7 percentage points.[14] The central question is how to convince people who do not think that science is a "fun puzzle" that it really is. The answer to that question, to the extent that there is one, seems to lie in the schools.

Changing Schools. A number of changes in educational programs have been suggested to make science education more female friendly, and some universities have gone to great lengths to make science more appealing to women. For example, the University of Illinois has established a program called Women in

Math, Science, and Engineering (WIMSE).[15] All-female dormitory floors for science students provide special computer labs, study lounges, and personal tutors. Special access to faculty is provided, including catered dinners with them. Other universities, including Purdue, the University of Wisconsin, and the University of Michigan, have established similar programs. An evaluation of the Michigan program recently found that women in the program were more likely to graduate with math and science degrees than were control groups of women not in the program and men, although method of selection into the program could bias the result.

The provision of educational and support resources to females that are not available to males may well be impermissible under Title IX of the Educational Amendments of 1972, which prohibits sex discrimination in education. Beyond questions of legality, however, there is a more fundamental question of whether female scientists are "worth more" to society than male scientists to justify the greater cost of producing them. It is not at all surprising that devoting more resources toward a particular group of students will increase their likelihood of success. Provision of tutors, computer resources, and personal faculty attention would probably increase the persistence and performance of some male students as well. To the extent that the programs are justified by the nation's need for scientists and engineers, a more broadly applied program would be more appropriate, but it might not provide the egalitarian outcome desired.

Some advocate making science faculty more open and accessible to students, providing a more "caring approach." Thus, we are told that it is not enough for faculty members to provide good lectures and conduct world-class research, they must also provide a "welcoming and supportive environment" for students.[16] However, many individuals who do world-class research may not be warm and caring and may be incapable of pretending to be so. As David Lubinski has observed, "In part because of the intensity with which these individuals approach their work, the highly creative, as a group, are also known to be difficult in interpersonal relationships, socially harsh, and abrasive."[17] Thus, it may be expecting too much to expect universities to go much beyond rhetoric in this respect. Would we expect—or even want—a major research university to decline to hire a Nobel-quality scientist in favor of someone less brilliant but more personable?

Many of the suggestions to increase female interest in the sciences are quite patronizing to women. Some believe that such things as providing pizza parties for physics students,[18] creating public service announcements "that use the words 'engineering' and 'fun' in the same sentence,"[19] or naming the engineering school after a woman, as the Rochester Institute of Technology has done, are at least part of the answer. The idea that women are going to make fundamental life decisions on the basis of such trivialities does not say much about their advocates' respect for women's autonomy or judgment, and it is unlikely that such initiatives will have a long-term impact on the number of women in science.

Changing Science. The remedy of last resort—if women cannot be convinced that science is fun and rewarding and if universities cannot attract sufficient women—is to change science itself. Since "we can't fix the girls, we have to fix science," says one scholar who has written extensively about women in science.[20] Declares another, "We really have to rethink our whole notion of what science is and how it functions."

The notion that if a discipline is less attractive to women than to men, then it is flawed and should be changed is nothing short of bizarre. The discipline is the discipline: physics is physics, not social psychology; engineering is engineering, not nursing. Perhaps there is a way to characterize bosons and quarks in social terms and to describe their behavior nonmathematically so that the nonmathematical will not be disadvantaged, but it seems doubtful.

It is not clear what could be done to "change science" to make it more female friendly, and if changes were simultaneously to make science less friendly to the men and women who currently comprise the scientific community, society would surely be the poorer for the change. To the extent that many women avoid certain scientific fields because the fields require a high level of mathematical ability, an intense investigative and theoretical orientation, and a tolerance for a somewhat socially isolated existence, the question is how to change those aspects of science without diminishing science in the process. How can mathematics, physics, and engineering be converted into fields that require less mathematical ability, especially given the increasingly mathematical nature of the sciences?[21]

It is worth pondering just exactly why it is important to change science if that is the only way to attract more women. The argument distills down to the proposition that society is somehow better off if there is proportional representation of the sexes in every subfield. Is this because women will bring some superior insight to particle physics or mechanical engineering? Will "women's way of knowing" yield a safer bridge or lead to discovery of a new fundamental particle? Although it would seem unlikely, the president of the National Academy of Engineers has asserted that "to the extent that engineering lacks diversity, it is impoverished. It is not able to engineer as well as it could."[22] One waits in vain for data to support this claim.

Female scientists have, of course, made not only important, but also distinctive, contributions to their fields. The fields that have clearly benefited from a "woman's perspective" include cultural anthropology, primatology, and psychology. However, as Sarah Hrdy has suggested, this is not so much because women "do science" differently from men, but because they ask different questions to start with.[23] Although not universally the case, male anthropologists, primatologists, and psychologists tended, not surprisingly, to focus on male activities, both because they found them more interesting and, especially for the anthropologists, more accessible. Women in these fields have made important contributions that men were simply much less likely to make, but they are not fields where women are currently underrepresented, as a majority of degrees in these fields already go to women.

An inherent self-contradiction in the argument for "diversity" is often overlooked. Many take the disparities in representation of men and women as proof that the deck is stacked against women, on the rationale that we should expect women to sort themselves into jobs in the same proportion as men. Yet, we are simultaneously told that women bring a different perspective and different values to the workplace. If men and women truly do have different perspectives and values, however, then why on earth would one expect them to have the same occupational preferences?

Concern for statistical disparities seems a trifle one-sided. Much concern is expressed that only 35 percent of bachelor's degrees in physical sciences go to women but little over the fact that only 27 percent of bachelor's degrees in psychology go to men. The fact that only 41 percent of Ph.D.'s in biology go to women is viewed as a problem, but the fact that only 42 percent of Ph.D.'s in anthropology go to men is not.[24] Much attention is spent thinking about how to make scientific professions more "female friendly," but little is devoted to making professions such as nursing more "male friendly." It is difficult to imagine someone saying, for example, "We need to completely rethink the profession of nursing; it is not attracting enough men; since we can't fix men, we need to fix the nursing profession." The assumption that men do not become nurses because they do not want to seems to be sufficient.

Increasing the Number of Women in Blue-Collar Jobs

Arguments for greater integration of blue-collar jobs often parallel those for scientific occupations. Again, however, the value of statistical parity in blue-collar jobs is not obvious, if the process leading to disparities is fair. Not only are the benefits difficult to identify, efforts to integrate some jobs create substantial risk.

The representation of women in firefighting is, as we have seen, very low—indeed less than 1 percent in some large metropolitan fire departments—a situation that has led to pressure to increase the numbers. Unlike the case with sexual integration of police forces, however, where there is a plausible argument that women are better than men at some aspects of the job—such as defusing potentially violent domestic situations—or racial integration in fire departments, where there is a plausible argument that residents of minority neighborhoods will be more cooperative toward a racially integrated department, no such argument exists for sexual integration of fire departments. No one argues that adding women will bring a needed perspective or allow fire departments to fight fires more effectively; nonetheless, efforts proceed apace.

The New York City Fire Department, for example, has engaged in an extensive recruiting campaign to attract female applicants. An advertising campaign to encourage women to apply is not troubling in itself, although one might question whether it is a wise expenditure of money at a time when thousands of applicants (albeit almost all men) are eager to become firefighters. The department went beyond just advertising, however. It took the additional step of inviting women who passed the preliminary written test to participate in a program of intensive physical conditioning with professional trainers to prepare them for the physical test, a test including such job-related tasks as raising ladders and handling hoses. The program was apparently a success, because nine of the eleven women (out of 105 female test takers) who scored sufficiently high to be hired had participated in this program, and more women than usual successfully completed the test. (In comparison, 57 percent of the men who took the test passed.)[25]

The training program providing special help to pass the test should be troubling to the residents of New York. Those candidates who could pass the test only after a period of intensive conditioning are likely to lose a good measure of their physical fitness after the test unless the conditioning continues. After the test has been passed, however, the department apparently stops providing the women with extra training. If passage of the test is as job related as the department maintains, perhaps it should require all firefighters, male and female, to requalify periodically. However, given that almost all of the women but probably far fewer of the men (given their relatively high pass rate) had engaged in extraordinary training for the test, one must especially question whether the female firefighters continue to maintain the strength necessary for the job. It is no consolation to a fire victim that the firefighter could have carried him to safety at the time she took her test if she cannot do so when his life depends on it.

Some of the assumptions underlying efforts to obtain proportional representation throughout the labor force should be reexamined. It may be that there are legitimate, even powerful, reasons for increasing the number of women in certain occupations. These reasons are likely to be specific to the particular occupation (such as women in policing), however, rather than applicable generally as a matter of principle to all occupations.

The Legal Treatment of Statistical Disparities

Although the antidiscrimination laws were not designed to require employers to achieve proportional representation within occupations by sex, doctrines of proof developed under those laws often have much the same effect. There are two basic theories of discrimination under Title VII of the Civil Rights Act of 1964, the flagship of federal antidiscrimination law. Under the *disparate-impact* theory, it is unlawful for an employer to use facially neutral selection practices that create an adverse impact on a protected group (race, color, sex, religion, and national origin), even if inadvertent, unless the employer can justify its use of the practice.[26] Under the *disparate-treatment* theory, it is generally unlawful for employers intentionally to treat individuals in a protected category differently from members of other groups.[27] The role of disparities differs under these two theories. Under the disparate-impact theory, statistical disparities resulting from discrete employment practices are presumptively unlawful as a matter of substantive law, while under the disparate-treatment theory, statistical disparities may be used as evidence that the employer has engaged in intentional discrimination.

The Disparate-Impact Theory. Statistical disparities are at the core of disparate-impact cases, in which the question is whether some employment practice falls more heavily on one group than another. For example, height-and-weight requirements tend to have a disparate impact against women, and standardized-tests and educational requirements often have such an impact against some minority groups.[28] Unless the employer can prove that these requirements are "job related for the position in question and consistent with business necessity,"[29] they are illegal irrespective of the employer's good faith in adopting them.

The tendency of disparate-impact theory to pressure employers toward proportional representation flows from the difficulty of justifying practices with a disparate impact. It is not enough for the employer to show that the practice is a rational one; rather, it must establish that the practice, as an empirical matter, is important to performance of the job, which, under current interpretation, often requires costly validation studies. Even if the employer can demonstrate that some factor, such as strength, is necessary to the job, it may be called upon to justify the particular cutoff point that it uses. For example, when the Philadelphia-area transit authority implemented an aerobic-capacity requirement for its transit police—who are often called upon to run for several blocks to provide assistance to other officers—the requirement was challenged on disparate-impact grounds. The trial court held that the transit authority's elaborate validation studies had satisfied its burden of justification. The appellate court reversed, however, on the ground that even if some relatively high level of aerobic capacity is necessary for the job, the employer had not proved that the precise cutoff that it chose reflected the "minimal qualifications" necessary to perform the job. Just because the employer can show that "more is better," said the court, does not establish that its cutoff was justifiable. Of course, any precise cutoff separating the "minimally qualified" from the "unqualified" is difficult to justify when the relationship between aerobic capacity and job performance is a linear one; any cutoff is necessarily somewhat arbitrary even if the value selected is a reasonable one. Nonetheless, the trial court on remand held that the employer had satisfied even that onerous standard, concluding that any lesser standard "would result in officers unable to arrive in a timely fashion to help a fellow officer in an assist or back-up, and officers unable to apprehend perpetrators."[30] As of this writing, the case is again on appeal, and it remains to be seen whether the practice will ultimately be upheld.

The difficulty of justifying practices with a disparate impact imposes substantial pressures on employers to achieve proportional representation by avoiding practices that lead to disproportionality. When employers abandon job-related practices because of their concern that they will not be able to prove their job-relatedness, efficiency and even safety may suffer. For example, after the physical test employed by the New York City Fire Department was struck down in 1982 because of the city's failure to persuade the trial judge that it was sufficiently predictive of job performance, the city eventually adopted another physical test that was ultimately found to be sufficiently job related (although its impact was just as great). Before the city's adoption of the latter test, however, it hired a number of female firefighters pursuant to a special procedure approved by the court. The standards employed were so much lower than those that were subsequently validated that of the thirty-six female firefighters on the force in the year 2000, thirty-one were hired pursuant to that 1982 test; in the following eighteen years, only five women were hired.[31]

The disparate-impact theory was initially adopted by the Supreme Court for the laudable purpose of eliminating "arbitrary" barriers to advancement of women and minorities. It is strong medicine, however, and unless judiciously administered it can impose substantial pressure on employers to engage in quota hiring.[32]

The Disparate-Treatment Theory. Statistical disparities are also central to most classwide claims of discrimination, in which the plaintiffs must show that discrimination is the employer's "standard operating procedure—the regular rather than the unusual practice."[33] Statistical evidence is often central in such cases, because if the employer is excluding some group on a systematic basis, that group will probably be statistically underrepresented in the employer's work force.

Statistical evidence of underrepresentation is admissible in classwide cases, according to the Supreme Court, because imbalances are often "a telltale sign of purposeful discrimination." It is appropriate to draw an inference of discrimination from statistical disparities, according to the Court, because "absent explanation, it is ordinarily to be expected that nondiscriminatory hiring practices will in time result in a work force more or less representative of the . . . composition of the population in the community from which employees are hired."[34] Admittedly, in those circumstances in which a job requires some specific qualification, like an engineering degree, the analysis focuses on the disparity between the employer's engineering work force and the proportion of women engineers. When no special qualification with a known statistical frequency is required, however, courts often simply compare the employer's work force with the general labor force.

The presumption that the work forces of all nondiscriminating employers would be sexually balanced—which is the "Central Assumption" underlying the use of statistics in discrimination cases[35]—is so at variance with reality that one must question a legal system that erects it, as should be apparent even to readers willing to accept only a fraction of the evidence in this book about the systematically different preferences and priorities of men and women. The assumption of equal interest is no more appropriate if one believes that the systematically different preferences of men and women arise from lifelong socialization, because the question in a discrimination case is whether the specific employer, not society, engaged in discrimination.

An example of the Central Assumption in action is the case of *Catlett v. Missouri Highway and Transportation Commission*, a class-action lawsuit alleging a pattern of sex discrimination in hiring for highway-maintenance positions.[36] These positions required an eighth-grade education and an ability to operate lightweight motor equipment. Duties included mowing along highways, plowing snow, filling potholes, maintaining rest areas, and servicing and repairing equipment. During the relevant time period, the employer had hired eight women and eighty-nine men. The plaintiffs had argued that based on the percentage of women in the general labor force, the Highway Department should have hired forty-six women, leaving a "shortfall" of thirty-eight women. Missouri, in contrast, argued that this statistical showing failed to account for the fact that women have less interest in these positions than men. The state argued that the proper comparison was between the rates at which male and female applicants were hired; among those who actually applied, the selection rate for women was marginally higher than that for men. Despite the fact that the individual claims of discrimination by the class representatives were rejected, the lower court nonetheless found that the Highway Department was liable for discriminating against the hypothetical thirty-eight women. The Court of Appeals affirmed on the ground that the defendant's argument of differential interest was mere "conjecture," despite the fact that women applied for the jobs at a rate less than 10 percent that of men. There is something wrong with a legal system that is based on an ideology so patently out of step with not only common sense but also a wealth of empirical data.

The conflict between the Central Assumption and an accurate understanding of human psychology is perhaps nowhere better—or more expensively—illustrated than in the litigation by the Equal Employment Opportunity Commission (EEOC) against retailer Sears, Roebuck and Company.[37] The EEOC sued Sears for sex discrimination in the hiring of commission salespersons. At trial, the EEOC introduced massive statistical evidence showing that women were underrepresented in these positions compared to Sears's noncommission sales force, a fact that Sears did not contest. Although large numbers of commission sales employees were women (over 30 percent in most categories for the years in question), the number was small compared with the percentage of its noncommission sales workers (approximately 75 percent) who were women. Despite the EEOC's abiding conviction that Sears was discriminating against women on a monumental scale, and despite the fact that its case against Sears involved 47,000 hiring and promotion decisions in 900 stores, the EEOC did not call a single woman to testify that she had been discriminatorily denied a position.

In defense, Sears presented evidence that commission sales jobs were significantly different from the noncommission sales jobs held predominantly by women, the latter primarily involving ringing up purchases on a cash register. Commission sales, in contrast, usually involved "big ticket" items; entailed financial risk because low sales mean low compensation and perhaps discharge;

required a high degree of technical knowledge; and often required irregular hours and visits to customers' homes. Sears also presented testimony from managers and experts to the effect that its affirmative-action efforts had met with only limited success because women were less interested in commission sales, in large part because of the "cut-throat competition" and the high degree of pressure and risk associated with the positions. The Seventh Circuit affirmed a judgment for Sears based on Sears's evidence of differential interest and qualifications coupled with the EEOC's failure to introduce evidence of "flesh and blood" victims of discrimination. Although Sears ultimately prevailed, it took 15 years of litigation, 135 days of trial involving 20,000 pages of transcripts, 49 witnesses, and 2,172 exhibits consisting of another 22,000 pages, not to mention, of course, tens of millions of dollars in attorneys' fees (and millions of tax dollars as well).

The *Sears* decision evoked hostile, if not hysterical, reaction among academics. One law professor expressed dismay that the court "has explicitly found evidence of women's lack of interest in a job relevant to a determination of whether the underrepresentation of women was caused by intentional discrimination,"[38] as if the level of women's interest is irrelevant to how many women hold the jobs. Others asserted that the "lack of interest defense has the potential to eviscerate Title VII's role in dismantling job segregation."[39] Yet another complained that the decision "enshrined gender stereotypes at the core of Title VII."[40] Feminist historian Rosalind Rosenberg was vilified for her testimony at trial, which had suggested, on historical grounds, why women may have different preferences from men.[41] Some would like to deny employers even this defense. Law professor Vicki Schultz, for example, has gone so far as to suggest that an employer's very willingness to raise the argument of differential interest should be used against it in a lawsuit as evidence of its discriminatory attitudes.[42]

A legal regime based upon an accurate understanding of the nature of men and women would not require an employer to spend fifteen years litigating the question whether women are less competitive and more risk averse than men. Despite the fact that Sears prevailed, future defendants faced with a similar claim will have to start from scratch as long as the counterfactual assumption of equal interest continues to be applied. In each case, the defendant will be required to find expert witnesses who are courageous enough to take a position that is contrary to current orthodoxy (especially in the academy) and trust that the trier of fact has an open mind.

Comparable Worth

A commonly urged remedy for the wage gap is to require equal pay for work of "comparable worth," a practice that sometimes goes by the more politically palatable but less descriptive term "pay equity." Comparable worth goes far beyond the Equal Pay Act's requirement of "equal pay for equal work," which prohibits sex differentials for very similar work. Instead, it requires equal pay for different, even dramatically different, jobs, as long as the jobs are believed to require "comparable skill, effort, and responsibility under similar working conditions."[43] Jobs are usually assigned points based upon these factors, a practice that allows comparison of very different jobs along a common dimension.

The impetus behind comparable-worth initiatives is the occupational segregation that is pervasive in the labor market. Although equal-pay requirements ensure that employers pay men and women the same wage when they do the same job, men and women, as we have seen, often do very different work. Equal-pay requirements therefore do little to close the wage gap. Comparable worth, in contrast, allows equalization of wages not only within jobs but also across jobs. Most comparable-worth initiatives do not apply comparable-worth principles across the board, however, but only between predominantly male and predominantly female jobs.

Proponents of comparable worth often act as if the only deviation of wages from comparable-worth principles occurs between male- and female-dominated occupations. They assume that these deviations result from sex bias. However, substantial "inequities" are revealed even between jobs having the same sex composition. For example, an Illinois job-evaluation study found that electricians (mostly male) were paid more than accountants (also mostly male), yet the study awarded 889 points to accountants and only 578 to electricians.[44] What accounts for the compensation differential? No doubt, supply and demand played a very important role. What accounts for the overwhelmingly higher point award to the lesser-paid accountants? Probably in large part the fact that job-evaluation methods, devised by white-collar consultants, tend to favor white-collar over blue-collar work, a fact responsible for much of the relative advantage that comparable-worth systems provide to women.

Supply and Demand

The notion that comparable jobs should be comparably paid has a certain superficial appeal. It sounds reasonably straightforward in application, and it has about it an aura of fairness. Who, after all, could quarrel with a requirement that employees be paid the same amount for comparable jobs? Most economists quarrel with it, as it turns out, because they are skeptical that substitution of a job-evaluation system for market forces is likely to benefit the economy in general or women in particular.

Comparable worth poses a host of problems, chief among which is the fact that it divorces the primary determinant of wages in a free market—the forces of supply and demand—from wage determination. How does one decide how much a job is "worth"? In a market-based system, the decision is made the same way one decides how much, say, water and wine or iron or gold are worth: they are worth what people are willing to pay for them. Gold and wine command much higher prices in the market than iron and water, but not because they are more "important." Life is impossible without water, merely unpleasant without wine, and modern civilization is unimaginable without iron, but simply less ornamental without gold. Given our dependence on water and iron, then, why are they so much cheaper than wine and gold? Primarily because the demand for wine and gold is higher *relative to supply* than is the case with the more abundant water and iron.

People today tend to be quite skeptical about governmental price determination of tangible commodities, understanding that in most cases the market does a better job of price setting than government does. They should be no less skeptical of centralized determination of the price of the principal commodity in today's economy—labor.

The Subjectivity of Job-Evaluation Systems

Once supply and demand no longer determine wages, something must take their place, presenting the puzzle of how to devise an alternative to the market that will work as well. The first practical hurdle is determining which jobs are comparable. How does one compare the working conditions of a nurse to those of a garbage collector, the necessary skills of an electrician to those of an accountant, or the effort required of an engineer to that required of a high-school teacher? Such determinations are highly subjective, and each job-evaluation system has its own method for making them. While job-evaluation scores have a patina of objectivity—they are numbers, after all—they are heavily value laden.

Even within a single job-evaluation system different evaluators may score jobs quite differently.[45] When the state of Illinois conducted a job-evaluation study of twenty-four positions, two sets of evaluators (using the same system) differed not only on the total number of points awarded to the jobs, but also on the relative rankings.[46] Although the two sets of figures were highly correlated ($r = 0.92$), if compensation were set on the basis of one set of figures, only 85 percent of the variance in compensation would be accounted for by the second set of scores, leaving a 15 percent unexplained disparity of the sort we have often seen attributed to invidious factors.

Application of different job-evaluation systems may produce even greater disparities. For example, Richard Burr found wide variation among states in rankings of Data Entry Operator, Laundry Worker I, and Secretary I. Minnesota ranked them 1–2–3, respectively, while Iowa ranked them 3–2–1; Vermont ranked them 2–1–3, while Washington ranked them 2–3–1. In Minnesota, registered nurses, chemists, and social workers were rated equally, while in Iowa nurses were worth 29 percent more than social workers, who were in turn worth 11 percent more than chemists. In Vermont, on the other hand, social workers were worth 10 percent more than nurses, who were worth 10 percent more than chemists.[47] As a Brookings Institution study observed, "Subjective evaluations enter at every step of the way: in determining what attributes to include in the job evaluation, in setting the point weights for each attribute, in deciding how many points each job should get for each attribute, and in calibrating the resulting point scores with pay."[48]

Job-evaluation studies can be easily manipulated. After Ohio rated public-sector jobs on a pay-for-points basis, the median hourly wage for women remained 13 percent less than that for men.[49] Rather than accepting the results as an objective indicator that men's jobs were more demanding than women's, the state concluded that the evaluation system itself must be biased and had to be modified. Similarly, when New Jersey conducted a wage study, it rejected the "working conditions" factor, apparently because most job-evaluation systems award no extra points for office work since offices are defined as the standard for "good working conditions." Because women work disproportionately in offices, extra payment for unpleasant working conditions would disproportionately benefit men.[50] New Jersey's initial study had found that women received only about 70 percent as much as men but that their pay per point was not lower; therefore, women were not deemed underpaid according to comparable-worth principles. The response was to revise the job-evaluation system to place more weight on factors that would favor women. Demonstrating that the purpose of the study was more to increase women's pay than to rationalize pay practices, the New Jersey report explained: "The increase in points for titles dominated by females and minorities met the goals of the Task Force to modify compensable factors to achieve pay equity."[51]

Even comparable-worth advocates acknowledge, at least in their more candid moments, their real purpose. Steven Rhoads quotes an academic at a conference of the National Committee on Pay Equity (a comparable-worth advocacy organization) as stating: "This is not social science . . . this is advocacy. . . . You can manipulate these things; you know, tell me the results you want, and we'll go get it, and if you don't think that's the case, I think you are being extremely naive."[52] Another comparable-worth proponent advised that, in order to ensure large raises, the various job-evaluation systems must be examined at the beginning to identify the one that will produce the desired results. "Every detail" of job evaluation is "political," she said, but it is important to "act as if [the details] are technical issues, because you're using [the system] to get pay equity gains."[53] "Pay equity," to these advocates, means increasing women's wages, not rationalizing compensation systems.

Comparable-Worth Systems Require Continual Fine-Tuning

Once jobs are assigned a point value, the work of developing and administering the system has just begun. Suppose an employer employs large numbers of nurses and a fairly small number of engineers, and the job-evaluation system reveals that the "market" has caused nurses to be underpaid and engineers to be overpaid. Nurses' wages may be increased and those of engineers reduced, although incumbent engineers may be "grandfathered" at the higher rate. Now the employer is paying above market price for nurses and below market price for new engineers. The employer should have no trouble attracting nurses, but how does it attract engineers at below-market wages? The only course apparently open to our employer if it wishes to continue employing both nurses and engineers is to return engineers' wages to the market rate to meet the competition and simultaneously raise the wages of nurses to the engineers' rate to satisfy comparable-worth principles. This may be a feasible option for public-sector employers, because their monopoly position may allow them systematically to pay above-market wages. Employers in competitive sectors are more constrained, however; for them, a more attractive option is likely to be to continue to pay one group the market rate and contract out the work of the other group.

Basic economic theory, in addition to evidence presented in prior chapters, suggests that many predominantly female occupations are low paying because they have features that are attractive to many women, and women seem to be willing to work for lower wages. As a result, the supply of labor is plentiful relative to demand. Increasing the wage associated with those jobs will increase the supply of workers desiring them. Yet, no system of comparable worth can make an employer hire employees. The employer faced with a mandate to pay above-market wages for librarians may decide to make do with fewer librarians rather than cut back on book purchases. The remaining librarians will be paid more, but they may have to work harder, and other librarians will be out of a job.

Comparable-worth systems require continual modification. Jobs change over time with changes in technology, products, and operating methods, necessitating reclassification of affected employees. A surplus of applicants created by newly raised pay may allow an employer to hire more highly qualified employees. The rational employer response to more capable employees is to assign them more complex tasks, but then their job would have to be reclassified, perhaps entitling them to another raise. Similarly, if an employer is faced with a shortage of qualified applicants for other jobs because of newly reduced pay, the employer may have to reduce the complexity of the job to allow the employees that it is able to hire to function at an adequate level. Once it does so, however, it must also reduce the pay even further (or increase the pay of similarly unskilled workers). Moreover, the sex composition of jobs can change, as well, so that a "balanced" job may become male dominated or female dominated, or vice versa, so that job comparisons that formerly raised no issue become problematic.

Experience with Comparable Worth

Proponents of comparable worth are inclined to label the foregoing objections—which are based on such "Econ 101" notions as supply and demand—theoretical and speculative. However, Steven Rhoads's meticulous study of the operation of comparable-worth systems in Minnesota, the United Kingdom, and Australia revealed that virtually every warning of comparable-worth opponents has proved valid.

Minnesota's system, according to Rhoads, has resulted in budget increases, layoffs, lawsuits, and difficulty recruiting and retaining employees. The job-evaluation system has resulted in pay compression, leading employees to refuse promotions because the increased responsibility and stress are inadequately compensated. Low morale has been another problem, with those disadvantaged by the system predictably feeling bitter, and those benefited thinking they should have benefited more. Police officers and firefighters were outraged that secretaries were awarded "working conditions" points comparable to their own; secretaries, it seems, face "the stress of working between four walls" and of "meeting deadlines."

Minnesota's system had numerous undesirable effects. Librarians obtained raises of 20 to 60 percent, despite the dozens of applicants for every open position at the pre–comparable-worth salary, but nurses fared more poorly. While market forces were driving up the salaries of nurses in the private sector, the comparable-worth system—which was designed to insulate wages from the forces of supply and demand and in fact did so—left public-sector nurses substantially underpaid in comparison to the market. Although the law permits special adjustments to deal with shortages, the standards are onerous, and adjustments lead to substantial upward pressures for other comparably rated job categories. Public-sector child-care facilities, faced with increasing pay obligations for child-care workers, raised fees, cut staff, or increased the number of children taken without adding staff. The comparable-worth system declared computer systems programmers and other technical employees overpaid, although they were underpaid compared to the private sector, making it difficult to attract and retain competent and experienced people. Because of the significance of the sex composition of jobs under comparable worth, hiring decisions were made with an eye toward the effect on a job's classification as predominantly male, predominantly female, or balanced, a practice that, although probably illegal, was quite rational on the employers' part.[54]

The Effect on the Wage Gap

If gains under a comparable-worth system are a product of struggle rather than principle, there is little reason to think that women will be long-term winners. We have seen throughout this book that men care more about status and money than women and that they are more likely to engage in competitive behaviors to obtain them. As Rhoads has observed, comparable worth in the public sector has not usually caused men to have their wages cut and has been disproportionately influenced by feminist activists. In the private sector, Rhoads predicts, men will have much greater influence. Men have more to lose there, because competitive pressures acting on private-sector employers mean that wage setting is more nearly zero-sum, and feminist political advocates will likely not be able to exert as much influence in the private sector as in the public sector.

Even if a coherent system of comparable worth were possible, its effects on closing the wage gap would be limited, and it would likely have other negative effects on women. Comparable worth would have only a modest effect on the gender gap because women tend not only to work in different occupations than men but also to work for different employers. No matter what opinion one might have of the relative worth of textile workers versus road-construction crews, comparable worth would have little impact in equalizing wages of those two jobs, since few employers hire employees in both categories. Comparable worth is also likely to exacerbate occupational segregation, because it will remove the economic incentive for women to move into traditionally male jobs. Although one might have supposed that increasing the wages of traditionally female jobs would attract men into those jobs, that does not typically happen.[55] Moreover, because one major goal of comparable-worth advocates is to increase the pay of child-care workers, adoption of comparable worth would tend to work against their other goal of affordable day care.

Mainstream economists of all political stripes oppose comparable worth because they believe that economic growth is more likely to result from a labor market that pays in accordance with the principles of supply and demand than according to the dictates of comparable-worth consultants, administrative tribunals, judges, or juries. Economic growth rather than central planning, they believe, is the key to improving wages, as Australia's history of centralized wage determination and abysmal economic performance suggests. As the politically liberal economist Mark Killingsworth has stated: "Comparable worth does not recommend itself to liberals. Comparable worth is for silly people, not liberal people."[56]

Other Pay-Equity Measures: The Death of Merit?

There is probably little political support for a requirement that employers pay men and women the same amount without regard to the kind of work they do or how productive they are. Covert strategies with the same purpose and effect, however, have been underway for some time in the form of "pay equity" measures and discrimination remedies.[57] Because they tend to be couched in remedial terms, these measures are often not understood to be the sharp departures from meritocratic principles that they are.

Many academic institutions have conducted "pay equity" studies to determine whether male and female faculty are paid equally, making "pay equity adjustments" when they are not. Virginia Commonwealth University, for example, conducted a study to determine the causes of a ten-thousand-dollar difference in average salaries of male and female faculty. After controlling for doctoral degree, academic rank, tenure status, and number of years since beginning teaching, a residual of somewhat less than two thousand dollars remained. The study's design automatically attributed the remaining disparity to sex. The university then allocated almost half a million dollars to increase female faculty salaries.[58]

Male faculty members brought a sex-discrimination lawsuit. Under affirmative-action case law, the sex-based salary adjustment was valid only if the university had a firm basis for believing that the discrepancy being corrected was due to sex rather than to nondiscriminatory factors. The male plaintiffs argued that the university's salary regression was fatally flawed because it did not include many of the legitimate factors that the university actually used in setting compensation under its merit-pay system: teaching load, teaching quality, quantity and quality of research and publications, and service to the community. The university contended that these variables were "too subjective" to include in its statistical model—though apparently not too subjective to use in actually setting salaries—and that it was reasonable to assume equal productivity of the two sexes. The trial court granted summary judgment to the university, which means that it won without having to go to trial. A three-judge panel of the Court of Appeals for the Fourth Circuit upheld the university's position, ruling that the salary study provided an adequate basis to conclude that the discrepancies were attributable to sex.

The entire Fourth Circuit Court of Appeals reviewed the panel's decision and, in a 9–5 decision, overturned the district court's ruling in favor of the university. The majority ruled that the plaintiffs had shown that there was a factual dispute over whether the salary discrepancies were attributable to sex that would have to be resolved at trial. Judge J. Michael Luttig concurred in the judgment but went even farther. He expressed the self-evidently correct view that a study of a merit-based pay system that omits the primary measures of merit is so flawed as to be inadmissible. Therefore, he asserted, the university had *no* basis for believing that the disparities were due to discrimination and the plaintiffs should be entitled to judgment without even going to trial.

The majority's view that the university's salary study was deficient as proof of discrimination is so obviously correct that it is hard to imagine that anyone could disagree with it. Nonetheless, the trial court and the appellate panel had taken the opposite view, as did the five judges who dissented from the full court's opinion. The dissent contended that the university's assumption of equal productivity of male and female faculty was a reasonable one. However, as Judge Luttig had pointed out, the assumption of equal productivity cannot be a reasonable one when the question under investigation is precisely whether the salary disparities are due to productivity differences or to sex; to assume equal productivity is to assume discrimination. In any event, the plaintiffs had introduced evidence that the assumption of equal productivity was unwarranted, and they could easily have introduced even more, since, as we have seen, virtually all studies of scholarly productivity show male faculty, on average, to be substantially more productive than female faculty. Nonetheless, the dissent was offended by the majority's even "tacit" acceptance of the argument that the assumption of equal productivity was inappropriate, suggesting that "we should be far beyond that point today."[59] Notice that the dissent did not merely start with a presumption of equal productivity. Instead, the dissenters refused as a matter of principle to recognize the relevance of contrary evidence.

Even more striking is a passage in the dissent concerning the effect of sex differences in performance. It stated that "even if performance factors could measure and did in fact show differences between the productivity of men and women on the average, the only appropriate conclusion to be drawn is that performance factors improperly favor one sex over the other, not that one sex is actually more productive than the other." Suggesting that this breathtaking statement was not merely inartful drafting, the dissent went on to say that "as a matter of law" use of performance measures that lead to unequal salaries is illegal sex discrimination.

Adoption of the view of the dissent (or the many educational institutions that have adopted pay-equity plans) would result in an effective dismantling of merit-pay systems. The resulting practice is the incoherent one of paying on the basis of merit, which is by nature an inegalitarian form of compensation, unless it yields inegalitarian results. Of course, the inegalitarian results are reversed only selectively, since men who come out poorly in the merit system are generally without recourse.

A recent episode at MIT revealed that allegations of discrimination are easy to make and that confessions of discrimination and provision of remedies are seemingly just as easy. MIT made national headlines with its report *A Study on the Status of Women Faculty in Science at MIT*,[60] which concluded that senior women faculty faced systemic, though unintentional, discrimination. The report was described by the dean of the School of Science as "data driven," and it was widely praised.[61] The fact that the report actually contained very little data—certainly none showing discrimination—and the fact that the committee evaluating the question consisted primarily of the women alleged to be the victims of discrimination did not tarnish the national luster of the findings. The chair of the committee, a biologist whose complaints had triggered the investigation, was fêted at the White House and benefited in more tangible ways as well—a generous pay raise, an endowed chair, and a tripling of her lab space. The MIT dean soon thereafter accepted a university presidency.

Like Gertrude Stein's Oakland, however, there was very little "there there." Psychologist Judith Kleinfeld conducted a careful review of the methodology of the study and concluded that the report was not a scientific report but an "ideological tract," not science but "junk science."[62] Although the report had concluded that female faculty were victims of salary discrimination, the committee did not even have access to salary data. Even if there were a salary disparity between male and female faculty, might not there be legitimate nondiscriminatory reasons—such as differences in productivity—for any differences that might have existed? The MIT committee did not think so, although it did not collect any relevant data; indeed, it challenged the very legitimacy of such an inquiry. It is, the committee declared, the "last refuge of a bigot" to suggest that victims of discrimination "deserve it because they are less good."

Fortunately, not everyone is intimidated by groundless accusations of bigotry. Behavioral scientist Patricia Hausman and psychologist and statistician James Steiger conducted an empirical study of productivity of the MIT biology faculty.[63] They found a dramatic difference between senior male and female faculty. Three of six men but only one of five women had published more than one hundred papers between 1989 and 2000. Only one man, but the remaining four women, had published fewer than fifty papers in that same time. Three of the six men had more than ten thousand citations to their work, which was more than three times as many as the most commonly cited woman. Four of the men had more citations per paper than any of the women. Given these gross disparities, discrimination is not the most obvious explanation for any differences in salaries or lab space that might actually exist. But, like VCU before it, MIT assumed that it was and, in the process, called into question the legitimacy of rewarding productivity.

The policies described in this chapter are oriented primarily toward equalizing various statistical measures of workplace outcomes, such as representation in particular fields and wages. Because part of the explanation for sex differences in workplace outcomes rests on the domestic division of labor, a number of policies have been suggested that would lessen workplace/family conflicts, a subject to which we now turn.

12 Mitigating Work/Family Conflict

I found myself completely changed after the first child. Before, I was very interested in equality, but now I feel that there is a special tie between mother and child. I find myself giving priority to family over work. There might not be any noticeable objective effects of staying home on my career, but I am subjectively different. I am not as psychologically involved.
—female Swedish physician

Many believe that the primary obstacle facing women in the workplace is the conflict between work and family. Some therefore advocate an equal domestic division of labor between husband and wife, changes in work rules to make it easier and less costly for women to cut back at work—such as requirements that employers provide part-time work and limits on the number of hours that can be worked—and social welfare programs, such as subsidized day care and paid parental leave.

Although different surveys yield widely varying estimates of the sex disparity in household labor, husbands clearly do less than 50 percent. Men's hours of household work are relatively constant across marital statuses, whereas women's increase dramatically upon marriage. Although many women would prefer that their husbands do more housework, only a minority believe that the current allocation is unfair, and overt conflict over housework is relatively infrequent. Wives tend to be more leery of greater contributions by their husbands in child care, however, whether out of skepticism about their husbands' competence or a desire to preserve their "turf." Whether the current allocation of housework within couples is fair or not, however, it seems beyond the influence of public policy.

Expanding available part-time work would be welcomed by many women, but it would probably have little effect on the glass ceiling or the gender gap. Although it would maintain the labor-force attachment of some women who would otherwise leave the work force altogether, it would attenuate the attachment of a probably larger group of women who would shift from full-time to part-time work in line with the desires of a majority of women.

Limiting the number of hours that employees can work would reduce the gender gap, but only because it would result in greater compression of earnings by reducing the compensation of high earners. However, such limitations would also result in salary cuts that would make it more difficult for a couple to manage on a single income.

Availability of subsidized day care would also have little effect on the glass ceiling because female potential CEOs who stay home are not impaired by their inabil-

ity to purchase day care but by their unwillingness to delegate control of children's lives to others. Subsidized day care may actually increase the gender gap in compensation, because it would bring many low-skilled women into the workplace, and it may increase occupational segregation through creation of a large number of jobs that will be held primarily by women.

Provision of paid parental leave would likewise be a mixed blessing in terms of the glass ceiling and gender gap. Short leaves tend to maintain the labor-force attachment of women after they give birth, but longer leaves tend to discourage reentry into the work force. Although some advocate expansion of parental leave to increase the involvement of fathers with their children, expanded leaves are unlikely to cause a radical transformation in the allocation of parental care responsibilities.

Many students of the workplace believe that the primary impediment to full workplace equality for women is not discrimination by employers but rather the division of labor at home. Time spent on housework and child care is time that cannot be spent in the labor market. Recommended solutions to the work/family conflict are of three major kinds. First, husbands should pick up their fair share, equalizing the domestic burden. Second, employment rules should be modified to reduce the penalty on women who cut back on their workplace commitment, such as imposing maximum-hours limitations or requiring employers to provide more and better part-time work. Third, social welfare programs should be expanded to include publicly subsidized day care and paid parental leave.

Some of these suggestions, such as modification of the domestic division of labor, seem largely beyond the reach of public policy. Others, such as maximum-hours requirements, may benefit women who are now at a competitive disadvantage because they are not willing to work as much as men, but they are likely to impose costs on other women, especially those who would stay home if their husbands could earn enough to permit it. Still other initiatives, such as expansion of day care, may be good ideas for other reasons, but they are unlikely to have a substantial effect on the statistical measures of inequality that have been the focus of this book.

Allocation of Domestic Responsibilities

An equal division of household labor is seen by many as a precondition to equality in the workplace. Only when the imbalance in domestic labor is eliminated, the argument goes, can women hope to compete on an equal footing with men.[1] Men should do their "fair share" of housework—which is usually assumed to be 50 percent—and they should take equal responsibility for the day-to-day care of children.

Women even in dual-career couples perform more household labor than men, although the extent of the disparity between the sexes is very difficult to gauge because of the unreliability of the data. Most studies of time allocated to housework, child care, and market work rely on post hoc self-reports, answering questions of the kind "during the last 24 hours/week/year, how many hours did you spend doing housework/taking care of your children/working at your job?" Accounts in the popular press of the decline of leisure time are based primarily upon such reports. These studies show an increase over the past few decades in the total amount of work (household plus market) of both men and women.[2]

An alternative method of measuring how time is spent is the use of time diaries, in which subjects make contemporaneous records of their activities. Diary studies are believed to be substantially more reliable than after-the-fact reports, but they are also more difficult and expensive to perform.[3] Diary measures reveal that for most categories of time expenditure post hoc reports overestimate the time spent on reported activities. According to diary measures, for example, total work declined, for both men and women, between the 1960s and the 1980s, with the decline being sharper for women; in contrast, retrospective estimates for that period continued to rise.[4] Depending upon whom you believe, then, we either have less leisure time or more leisure time than we did thirty years ago.

The magnitude of the disparity between retrospective self-reports and time-diary measures can be substantial. Estimates of hours of market labor performed substantially overstate work hours; indeed, some overreport by as much as 50 percent.[5] Moreover, women overreport to a greater extent than men do; in one large study, women overreported more than men by five to eight hours per week,[6] although more recent figures suggest that the sex difference in overreporting may be vanishing.[7] One reason for disparities between diaries and retrospective reports is that people are increasingly likely to engage in personal activities at work. For example, almost half of all online shopping in the 2000 holiday season took place from work.[8] Time diaries are more likely to reveal the personal, rather than work, nature of time spent paying bills, arranging doctors' appointments, talking to the children, and making vacation arrangements.[9]

Retrospective reports of domestic activities also tend to be inaccurate. For example, women report three times as much time caring for children as is recorded in time diaries.[10] Husbands and wives both overestimate their time spent on housework tasks, with one study finding husbands overestimating by 5.8 hours and wives by 12.8 hours per week.[11] On the other hand, estimates of time spent on home repair are only half as large as time diaries reveal.[12]

The pattern of inaccuracy—with women overreporting both their household and labor market work to a greater extent than men, and the underreporting of home repair hours (work that tends to be performed by men)—means that reported sexual disparities in combined home and market work must be taken with a grain of salt. A number of studies have found that men and women have relatively similar hours when paid and household labor are combined.[13] Thus, the popular notion that men and women typically come home from forty-hour-per-week jobs and then the husband relaxes while the wife starts her "second shift" [14] is a gross distortion. While that pattern may describe some households, it certainly does not describe the average household, any more than does the pattern where only the husband is employed and the wife does very little housework, either because she is lazy and leaves it undone or because her husband's earnings pay for domestic help. Nonetheless, by any measure, women do more housework than men do,[15] even if the disparity may be less than often portrayed. Moreover, the weight of the evidence suggests that time devoted to housework has a negative effect on wages and that this effect may be greater for women, although studies sometimes produce conflicting results.[16]

Housework

A variety of explanations have been suggested for why men do less housework than their wives, ranging from sex-role socialization to spite. A study by Julie Brines found that "the more a husband relies on his wife for economic support, the *less* housework he does," [17] but a study by Joni Hersch and Leslie Stratton found that "the husband's share of housework time is significantly lower when he contributes a greater share of labor income." [18] The explanation for the former finding is that "striking" on housework is a way of asserting a threatened masculinity. The explanation for the latter is that when men earn more they have a greater comparative advantage in the labor market and greater bargaining power in the relationship. Perhaps there is something to one or both of these hypotheses, but there is an additional explanation for why men do less housework: they care less about it than women do.

Much of the literature comparing men's and women's contributions to household work starts from the unstated assumptions that men and women equally value the fruits of household labor and that men and women find such labor equally distasteful. Viewed in this light, the sexual disparity appears inequitable, especially when sex differences in market labor are ignored. But it is not clear that either of these assumptions is correct.

If men and women cared equally about housework being performed, one would expect that men and women would perform equal amounts of housework before marriage and that when the marriage terminates through divorce or spousal death, equality would be restored. That is not the pattern, however. Instead, a study by Scott South and Glenna Spitze shows that men's housework hours are relatively constant across various marital statuses and living arrangements,[19] but women's hours vary significantly. Although married men averaged one hour of housework per week less than single men, married women averaged twelve hours per week more than single women. Even before marriage, however, women did approximately one-third more housework than men. On the other hand, for the entire sample, men averaged fourteen hours per week more market labor than women.

South and Spitze's results suggest that the husband is not saying, "Now that I have a wife, I don't have to do housework," but rather something more like

"Now that I have a wife, I shouldn't have to do more housework than I did when I was single." A study of new parents found that although men responded to pressure from their wives to participate in household work, "they did not share their spouses' views of the importance of such work."[20] This difference in valuation of housework creates an additional variable in the "bargaining" that Hersch and Stratton hypothesize: all else being equal, the person who cares more about the house being clean is going to do the majority of the work.

Women are often quoted as saying that they want their husbands to do their part helping around the house. Strict egalitarians would object, saying that "helping" is not the correct word: the husband is not helping the wife; he is doing his equitable share as a partner in the enterprise. But for a large number of women, "helping" is the correct word. Economist Jennifer Roback explains: "When I want my husband to do 'his half' of household chores, what I really want is for him to do half of everything on my list of important things. But he has his own list. He values some things that I do not; he does not value all the things that I do."[21] She did not add that she not only wants him to do the things on her list, but that she also wants him to do them her way, but many women would. As Ellen Galinsky, head of the Work and Family Institute, has observed, "The wife is saying 'Help me. But do it my way.'"[22] Indeed, men often report that they might as well not do housework because they will get as much criticism for the way they did it as they will for not doing it. Myra Ferree found that husbands who expect their housework to be criticized do less housework and that the more the wife cares about having a clean house, the fewer conventionally female chores her husband does.[23] Why is it that women tend to want the housework done their way? Apparently because, just as many men like to feel that they are the breadwinner of the family, many women like to feel that they run the house.

While housework is viewed as degrading by many feminists, at least if it is performed by women, it can also, as South and Spitze point out, "express love and care, particularly for women."[24] Moreover, despite the asymmetry in domestic work, studies suggest that most women believe the allocation is "fair" and that overt conflict over housework is relatively uncommon.[25] In fact, many women actually enjoy domestic labor. On the Strong Interest Inventory, some of the largest sex differences are found on such tasks as cooking, sewing, and "home economics."[26] (This is not to say, however, that there are many women who enjoy such tasks as cleaning toilets.) Again, it is precisely women's lack of feeling of victimization that Deborah Rhode has labeled the "'no-problem' problem."[27]

At bottom, regardless of the respective allocations of household and market labor, and regardless of whether members of particular couples are happy about it, allocation of domestic responsibilities seems largely immune to policy manipulation. The notion that the state would intervene to require "the complete and equal sharing of both paid and unpaid labor," as advocated by Susan Moller Okin,[28] is unthinkable in a free society, and it is not an outcome that a majority of either men or women would endorse anyway.

Parental Care

The other realm of domestic "labor" in which husbands are often called upon to enhance their contributions is child care. Men's time spent in child care has increased more than their time spent on housework in recent years, but men's contribution is still seen by some as inadequate. It is more difficult to measure child-care hours than housework hours, however, because child care is often done contemporaneously with other activities. When Junior is working a puzzle on the floor while both parents read the newspaper nearby, is this child-care time or leisure time for the parents? If the parents are vacuuming or repairing a lamp at the same time that Junior is playing, is this household labor or child care? The ambiguity presented by such questions probably accounts for the 200 percent overestimation of child-care time found by Juster and Stafford. One time study was forced to eliminate child care as a category, because, when it was included, the time reported exceeded the available hours in a day (assuming six hours per night of sleep).[29]

Another question that arises is determining what child-care activities count as "labor" at all. Most would probably at least include diaper changing in this category. But what about reading to the child, playing games, taking trips to the zoo, and even feeding? What about school activities, supervising the scout troop, helping with homework, and walks in the park? These activities can be construed as labor, since one would typically have to pay strangers to induce them to perform these functions. However, the opportunity to perform this "unpaid labor" is precisely what causes so many women to withdraw, in whole or in part, from the labor force upon the birth of a child, and these activities are widely perceived as part of the joys of parenthood.

Notwithstanding these ambiguities, we are faced with calls for "a fundamental reorganization of parenting, so that primary parenting is shared between men and women"[30] by psychoanalyst Nancy Chodorow and declarations by law professor Ann Scales that "the abdication of fatherly responsibility on any level can no longer be tolerated."[31] One might think that before embarking on a campaign of rearranging family life we would have evidence either that parents are dissatisfied with the current asymmetry, or, if parents are not dissatisfied, that the asymmetry is harmful to children. There is not much evidence of either.

There does not appear to be a widespread desire on the part of women to have husbands become equal partners in caring for their children. Indeed, most women would not welcome such a change, at least in part because they do not think their husbands are as good at child care as they are, and they want to protect their "turf."[32] As Linda Haas has observed, "Despite popular rhetoric, women typically exhibit little interest in the concept of equal parenthood and . . . they have difficulty giving up responsibility for child care."[33] If the justification for a radical reorganization of household duties is not that it is based on the preferences of at least one of the parties to the relationship, an alternative rationale might be that it would be better for children if their parents shared the child-care responsibilities on a fifty-fifty basis. Yet there is little evidence for that proposition either.

Although abundant evidence shows that paternal involvement benefits children,[34] there is little evidence to show that the beneficial effects are limited to, or maximized at, a fifty-fifty division of parental care. Consequently, a showing that paternal involvement is beneficial does not necessarily mean, as some would have it, that more is necessarily better. Although many people seem to think that the ideal model of egalitarian parenting entails two parents who are largely fungible—usually along the feminine line—there is evidence to suggest that children benefit more from having a mother and a father than they do from having two mothers, one of each sex.[35]

Like increasing the husband's share of housework, altering married couples' internal allocation of child-care responsibilities seems beyond the reach of most likely policy initiatives. One possible exception may be parental leave, described more fully below, which has been justified both as a way to involve fathers more in the lives of their children and as a way to improve the position of women in the labor force. As we will see, however, it is likely that even with generous parental leave the sexual imbalance in parental care will persist.

Is an "Egalitarian Marriage" Best for Everyone?

A strong strain in current discourse advocates what has been labeled "egalitarian marriage," with the husband and wife being equally committed to their careers, earning the same amount of money, and spending the same amount of time on child care and other domestic obligations.[36] The assumption that "egalitarian marriage" is best for everyone—man, woman, couple, children, and society[37]—is an ideological one that assumes that men and women are fundamentally the same, and therefore identity of inputs and outputs in both work and family is ideal and healthy for everyone. While these may be some peoples' preferences and are entitled to respect as such, it is not clear why these preferences are more deserving of societal imprimatur than other, less egalitarian and more traditional, preferences.

Over the past two decades, a large body of research has focused on the effect of dual-earner status on men's psychological well-being. A regular finding of these studies is that notwithstanding the obvious psychic benefits many wives get from working, dual-earner status often has negative effects on men's psychological well-being, as measured by levels of depression, self-esteem, and satisfaction with job, marriage, and life in general.[38] While one would expect that the "breadwinner" role would be less important to marital satisfaction among men with nontraditional sex-role attitudes, this does not appear to be the case.

Although the conventional explanation for men's negative reaction is that men feel that they are losing status to their wives, which might reflect traditional sex-role attitudes, in fact a major source of the negative effect on husbands is the ever-present status competition among men. Men in dual-earner families often feel themselves impaired in their careers by their wives' employment, as indeed they often are. Single-earner husbands tend to earn more than husbands in dual-earning couples, and among husbands with working wives, there is a negative correlation between the number of hours worked by a wife and the husband's earnings.[39] Therefore, what may appear beneficial to the couple, such as a man's sacrificing $20,000 per year in income so his wife can earn $30,000, may seem like a loss to him because of the lower position in the status hierarchy that he consequently occupies at work. It may be, therefore, that a dollar in earnings for the husband is worth more to many couples than a dollar earned by the wife, both because position in status hierarchies is more important for men than for women and because the wife gets more status gains from her husband's position than the husband gets from his wife's. It is also worth noting that while men in dual-earner families have lesser marital satisfaction and leisure satisfaction than men in single-earner families, they have even lower occupational satisfaction.

The idea that dual-career men's dissatisfaction is a result of workplace competition, and not just men's sense of inadequacy when the wife brings home some of the bacon, is buttressed by the finding of Sue Crowley that men in dual-earner couples who consider themselves *adequate* breadwinners suffered significantly diminished sense of well-being, while those men who viewed themselves as inadequate breadwinners did not. If the issue were perceived adequacy in comparison to one's wife, one would have expected the opposite result. As Sandra Stanley and her colleagues have concluded, "Whether through the direct effects of participation in child care or the indirect impact of the overload on an employed wife, a dual-earning household is not optimal for men of ambition."[40] As they note, "As long as men are diminished in the status hierarchy by wives' employment and family responsibilities, role change will be distressing even for, and perhaps especially for, the most able and educated of men."

The insight of Stanley and her colleagues is a critical one, because it suggests that the solution to the friction caused by dual-earner status is not simply to change attitudes about sex roles. Although they view men's anxiety as being "generated by the continued linkages between gender inequality and the larger stratification system," in fact both their analysis and their results suggest that men's unease may have very little to do with sex roles or sexual inequality. The inequality to which they attribute causal influence is actually inequality among men, specifically between single-earner men who can pursue their careers with relatively little constraint and dual-earner men who face pressure to make career compromises in favor of their wives. Thus, the adverse consequences cited by Stanley and colleagues—"avoidance of housework, inadequate parenting, domestic violence, and high divorce rates"—may be consequences not of men's resentment of women's career success, but of their resentment of the negative impact that their wives' careers have on their own career success. This is one more instance where what some might see as competition between males and females is really about competition between males.

If male status seeking rather than sex-role attitudes underlies the dissatisfaction of many men in dual-earner couples, it is not clear that there is any easy answer and perhaps there is no answer at all. Men (especially) are "hard-wired" to seek status, some more than others. It is not realistically possible to cause them not to care about their status relative to other men, nor is it clear that we should want them to.

Perhaps the greatest lesson to be learned from the foregoing discussion is the importance of maintaining the possibility of the single-earner traditional household. Men with the greatest ambition tend to marry women with more traditional attitudes, which allows them to pursue their goals in a single-minded fashion.[41] As discussed previously, policies intended to make the lives of dual-earners easier can also make it economically more difficult for couples to live on the earnings of a single member. Preserving the single-earner option is important not only to allow the preferences of a greater number of people to be realized but also because we should frankly acknowledge that many of the single-earners will make substantial contributions to society.

Modification of Work Rules to Reduce the Costs to Women Who Choose to Cut Back

If increasing the domestic contributions of husbands would reduce work/family conflict, it would do so by decreasing the demands on women at home, thus freeing time for labor-market activities. Another way to reduce this conflict is to decrease the demands on women at the workplace, freeing more time for domestic activities. Making it easier and more rewarding for women to work part-time is one way to do so. Another way is to impose maximum-hour restrictions on all employees, which would reduce the advantage currently enjoyed by workers (predominantly men) who are willing, or even eager, to work long hours.

Part-Time Work

Part-time work is the favored arrangement for at least half of all women.[42] Of those not now working part-time but who would like to, some are not now in the labor force, while some are working full-time. Indeed, it has been estimated that there are approximately four million "involuntary" full-time workers—those who are working full-time but would prefer to work part-time.[43]

The preference for part-time work extends beyond just women with child-care responsibilities. In 1995, 27 percent of married women with spouse present worked part-time, as did 27 percent of unmarried women (single, divorced, and widowed). For men, marital status has more of an effect, with 20 percent of unmarried men and only 7 percent of married men working part-time.[44]

Sociologist Catherine Hakim has pointed out that although women with dependent children have the highest rates of part-time work in Britain, women without dependent children still work part-time at about two-thirds the rate of women with children.[45]

One proposed solution to work-family conflict is to make part-time work both better and more widely available than it now is. Part-time work today is often routine and associated with lower wages, often with no benefits. Part-time professionals sometimes complain that they do not get the most coveted assignments and are less likely to be promoted than their full-time counterparts. It should be pointed out, however, that despite activists' complaints about "marginalized" part-time workers, women who work part-time report greater job satisfaction than women who work full-time.[46]

The Effect of Part-Time Work on Wages. The so-called wage penalty for part-time work is often misunderstood. The overall gap in hourly earnings between part-time and full-time workers is indeed large. In 1995, median *hourly* wages for part-time female workers were 74 percent of those earned by full-time female workers, while for men the figure was 55 percent. At first glance, these figures suggest that workers (especially men) suffer a substantial penalty for working part-time, but part-time and full-time work and workers are often very different. Part-time work tends to be concentrated in relatively low-skilled and low-paid occupations, and many part-time workers are young and have little work experience. More meaningful comparisons involve equivalent workers doing equivalent work. A large-scale study controlling for occupation and employee attributes such as age and education found that part-time workers had hourly earnings of approximately 92 percent those of full-time same-sex workers.[47] A similar study by Rebecca Blank actually found a wage premium on part-time work for women, especially women in professional, managerial, and technical occupations, but a wage penalty for male part-time workers.[48]

Differences in compensation for part-time work are likely to reflect productivity differences. Some researchers have suggested that productivity of part-time workers tends to be lower simply because of the nature of part-time work. Productivity tends to rise with more hours on the job, until fatigue or boredom sets in and productivity declines. The first and last periods of work are likely to entail some "gearing up" and "winding down," and these transitional periods tend to be a greater relative proportion of the workday for part-time workers. Also, although two workers occupy the same job classification, they may in fact have different duties.[49] The wage premium found by Rebecca Blank for part-time professional, managerial, and technical women may reflect the fact that employers are more willing to accommodate part-time work in these categories—where part-time work is unusual—for employees who have already proved themselves to be highly productive.

Legal Mandates. Many are concerned that part-time work is a dead-end "ghetto" of unfulfilling low-paid work that carries no benefits. Law professor Joan Williams advocates lawsuits against employers on two theories: first, for failing to provide part-time work at all, and second, for not paying part-time work proportionally as much in wages and benefits as full-time work. She also advocates amending the Fair Labor Standards Act to increase overtime compensation in order to discourage employers from requiring long hours. This is not the place for an extended discussion of the legal doctrine governing these issues, but it seems unlikely that current doctrine would recognize the lawsuits she envisions. Equal Pay Act lawsuits challenging the failure to pay part-time workers the same hourly wage as full-time workers would probably fail on the statutory principle that the Act does not prohibit compensation disparities that are based on a "factor other than sex." A distinction between part-time and full-time workers is a factor other than sex even if it correlates with sex. As to challenges for failure to permit part-time work, it is not clear that such failures are the kinds of practices that can be challenged under the disparate-impact theory of discrimination; even if they are, employers will often be able to justify them.

Williams's proposals are at war with each other. She advocates payment of part-time workers in proportion to the number of hours worked, yet she also recommends increasing the overtime premium from time-and-a-half to double-time, reducing the overtime threshold from forty hours per week to thirty-five, and narrowing the exemption from overtime requirements for employees who are classified as administrative, executive, or professional. These two mandates—paying in proportion to hours worked and increasing overtime premiums—are in obvious conflict. Williams describes a part-time lawyer who complains that she receives only 69 percent as much per hour as full-time lawyers earn. Such payment obviously conflicts with a proportional-pay requirement, which would be dollar-for-dollar, but it accords with a requirement of premium pay for overtime. For example, in a regime that required double-time for every hour over thirty-five, an employee who works thirty-five hours per week or less would earn 71 percent as much per hour as an employee who works sixty hours per week, suggesting that Williams's part-time lawyer was appropriately paid. Moreover, if her fringe benefits were not reduced by her part-time status, the lawyer's total compensation per hour would be substantially more than the 69 percent figure reported.

Requiring proportional pay increases the cost of part-time work, thereby creating disincentives for employers to permit it. An increased-pay requirement therefore leads to pressure to force employers to offer part-time work. Although any legal mandate to provide part-time work would undoubtedly allow an employer a "safe harbor" if it has sufficiently weighty reasons for not permitting it, whenever an employer made a judgment that a part-time arrangement would not work, its decision could be second-guessed in court. Although Williams points to the increased loyalty that employer accommodations engender, that loyalty comes from voluntary action on the part of the employer. An *entitlement* to part-time work is more likely to engender conflict than loyalty.

Any mandated benefit is likely to have costs associated with it, even to the workers one is intending to benefit. Requiring employers to provide part-time work at increased cost may result in the reduction of wages and benefits of others. Moreover, because these costs will be associated primarily with female employees, the change will increase the incentive for employers not to hire women. Although such a response would not be legal if it entailed a preference for male employees, there is evidence that such benefits do in fact reduce female employment.[50]

That employers may sometimes benefit by extending part-time opportunities more broadly does not mean that a legal mandate to do so is in the interests of either employers or employees. As noted by Williams, "one strong message of the existing work/family literature is that no single solution works best in all workplaces."[51] Legal requirements tend be of the "one size fits all" variety, and what works well in one workplace may not in another. As a long-term solution, it would probably be better to allow employers to adopt such plans voluntarily, as they are increasingly doing. Some employers' resistance may collapse when a valuable employee wants to change to part-time. If instead, the employer's early experiences are with marginal workers claiming an entitlement to part-time status, the employer is likely to become soured on the whole prospect. Also, employees who are allowed to change to part-time status as a matter of employer grace, rather than as a matter of right, have a greater incentive to make it work.

The Effect of Part-Time Work on Workplace Outcomes. More widely available part-time work would have only limited effect on the glass ceiling and the wage gap, and the effect would vary depending upon who takes advantage of it. The major positive effect is that it would maintain the labor-force attachment of women who otherwise would leave the labor force. Particularly if they are able to keep the same kind of job they had before they shifted to part-time status, they will retain their job skills, continue to acquire experience (albeit at a lower rate), and thus have a higher stock of human capital when they return to full-time work. Consequently, their earnings and prospects for promotion will likely be substantially higher than if they had withdrawn entirely from the labor force.

Just as part-time work will maintain the labor-force attachment of some women who would otherwise withdraw, it will attenuate the work-force attachment of many women who would otherwise remain in full-time employment. Women who shift to part-time rather than remaining in full-time employment will tend to be worse off in terms of both compensation and prospects for promotion. Many women will remain in part-time work far longer than they would have stayed out of the work force altogether, and some may stay in part-time work permanently. Again, that is not necessarily a bad thing, but those who measure women's well-being solely by their paychecks and job titles may view it as such.

Maximum-Hours Requirements

Many writers are critical of workplace environments that demand long hours. Because such workplaces are said to rely on "male norms," some argue that maximum-hours rules should be adopted. Sociologist Lois Bryson, for example, has argued for a six-hour working day for everyone, not just for parents, and "it should not be optional."[52] Others have argued for a reduction of the normal workweek from forty hours to thirty-five or even thirty-two.[53]

Some suggestions are not explicitly aimed at reducing the number of hours worked but are animated by similar motivations and would tend to have the same result. For example, a study of academic physicians suggested that some of the barriers to women's academic careers could be "easily modified," such as by eliminating "after-hours" meetings.[54] One's initial reaction to such a suggestion is "sure, why not?" After all, why can't the meetings be scheduled during "normal business hours"? The answer, of course, is that they could, but the time spent in those meetings is time that would otherwise be spent doing something else. Thus, the suggestion that meetings be conducted during normal business hours is implicitly a call for a reduction in hours: employees should leave at 5:00 rather than finishing their normal duties at 5:00 and then attending a meeting.

Who Benefits and Who Pays? Limitations on hours would benefit some who work in demanding professions. The typical large law firm, for example, probably effectively requires fifty to sixty hours of work per week, on average. Assume conservatively that the requirement is just fifty hours (including both billable and nonbillable time), and add in a half-hour commute each way and a half hour for lunch. Assuming no weekend work, that amounts to eleven and a half hours per day, with the lawyer leaving home at perhaps 7:30 a.m. and returning home at 7:00 p.m. That is a difficult schedule to maintain with young children. Many mothers scale back at their job, leave the work force for some period of time, or take a part-time job somewhere else, in each case postponing if not abandoning prospects for partnership. In contrast, a thirty-five-hour workweek would allow these mothers to be home by 4:00 p.m., for a 30 percent reduction in hours (or a 42 percent reduction from sixty hours). Since all other lawyers would be working a restricted schedule, the mothers would not be penalized, and everyone should be happy.

Well, perhaps not everyone. Proponents of reduced hours generally recognize that if the workweek is reduced, salaries would be reduced as well. The assumption that salaries would be reduced only in proportion to the reduction in hours is incorrect, however. Law firms run on the hours worked by their lawyers. If hours are cut by 30 percent, salaries would have to be cut by even more, because many of the firm's expenses, such as office space, support staff, and fringe benefits, would not be reduced in proportion to the decrease in hours. A reduction in hours increases the expenses per hour of attorney time and decreases profits. This is a different way of saying that firms make more profits on the last hour worked by a lawyer than on the first, which is one reason that employers pay less per hour to part-time attorneys.

Reduced pay in exchange for reduced hours would be ideal for some, but it would be undesirable to others. While we often hear of the demise of the "Ozzie and Harriet" style of family—breadwinner husband and stay-at-home wife—

it would be a mistake to think that it has disappeared. In 1999, 23 percent of married-couple families having children ages six to seventeen (but none younger) followed that pattern, and 37 percent of families with children under age six did so (and, of those mothers working, 23 percent and 30 percent, respectively, worked only part-time).[55] A reduction in the husband's salary by 30 to 50 percent would make it more difficult to support the family on one salary, which may require the wife to work as well. Thus, a change designed to enhance the options of the working wife may deprive the wife who would prefer not to work of her preferred choice.

Social Welfare Programs

Another class of initiatives consists of public subsidies for work and parenthood. Publicly funded day care, for example, subsidizes working parents by providing them child care that allows them to work. Parental leave, on the other hand, subsidizes working parents by paying them for not working while their children are young.

Day Care

Many writers have advocated publicly subsidized "quality, affordable day care," although an obvious tension exists between quality and affordability. There is reason for skepticism about its effects on either the glass ceiling or the gender gap in compensation, however. Lack of impact on statistical measures of equality is not by itself a reason to oppose creation of an entitlement to day care, because it would certainly benefit some people. If such a program is to be adopted, however, it should be for the right reasons.

Women who are potential chief executives or law firm partners—women who could contribute to "shattering the glass ceiling"—are not substantially impaired in their careers by a lack of public subsidy for day care. They are usually members of highly affluent dual-earner couples who are quite capable of paying for higher quality day care than the government is likely to provide. It is not the lack of day care that is typically the problem for them; it is their unwillingness to view their "children's needs as completely delegable."[56] Such women are not now completely free from "day-care problems," but government subsidies are unlikely to make them any freer.

Plentiful and inexpensive child care would also have little influence on the gender gap. In Britain, for example, statutory maternity leave and more widely available child care have had little effect on mothers' overall earnings.[57] Subsidized day care may actually *increase* the gender gap, depending upon who takes advantage of it. The most obvious groups who would benefit are low-skilled women who do not work because they cannot earn enough to make the cost of day care worthwhile and two-earner couples who must strain to pay for day care. Benefits to these groups would have little impact on either the glass ceiling or the gender gap.[58] Free universal day care has been predicted to increase women's labor-force participation by as much as ten percentage points. If the additional women have lower average earnings than women already in the work force, as seems likely, more widely available day care would actually widen the compensation gap.

Another likely effect of widely available day care is an increase in occupational segregation, as most day-care providers will continue to be women. One of the consequences of Sweden's broad entitlement to child care and other care services is that Sweden has the highest rate of occupational segregation in Europe, and one of the highest in the world, because most of the providers of such services are female.[59] Ninety-six percent of the day-care providers in Sweden are female, which is similar to the rate in the United States today. Almost 20 percent of nonagricultural female workers in Sweden are in one of the following occupations: child day-care center staff, social worker, municipal home help, and assistant nurse/attendant. Again, this sexual imbalance is not a reason not to offer the benefit, but those offended by statistical measures of "inequality" may judge it troubling.

Parental Leave

One common proposal to assist parents, especially mothers, is the expansion of parental leave. Under the federal Family and Medical Leave Act of 1993 (FMLA), eligible employees are entitled to twelve weeks of unpaid job-protected leave, during which the employer must continue health insurance coverage. Beyond this federal entitlement, parental leave is a matter of contract, employer policy, or state law (and only a few states have such requirements).

Many advocates of parental leave invoke the example of Europe. European countries vary widely in the extent of maternity and parental leave, although most are more generous than the United States.[60] Sweden provides fifteen months of paid parental leave to the couple, twelve months of which is at 80 percent of salary (with a cap), while the remainder is paid at a flat rate. Leave can be used at any time until the child's eighth birthday, though only one parent can take leave at a time. One month of the leave is nontransferable between parents, so the mother gets only fourteen months even if the father takes no leave. Employees have a statutory right to work just six hours per day (with a corresponding reduction in pay) until the child is eight years old.[61] Most European countries provide flat-rate payments, however, rather than payments tied to salary. While some countries provide longer leaves than Sweden, these leaves tend to carry much smaller remuneration. Germany, for example, provides three years from the birth of the child, but the monthly financial stipend is only 600 DM, and it runs for just two years.

Characteristics of Parental-Leave Plans. Parental-leave plans can take an almost infinite variety of forms. However, there are three primary issues that need to be resolved in adopting any plan: the length of the leave, whether and to what extent the leave is paid, and whether payment comes from the employer or the government.

The length of the leave affects both the burden of the leave on employers and the effects on employees. The longer the leave, the greater the potential burden on employers and the more likely the employer will have to attempt to hire replacement employees, not always an easy task at times when unemployment is low. Large employers with frequent turnover may have little difficulty in accommodating a returning employee without discharging the replacement, but small employers, or even larger employers where the particular job is unique, may have more difficulty managing the transition.

The longer the leave, the greater the workplace impact on employees eligible for the leave. Longer leaves are associated with greater depreciation of skills, so employees who take long leaves suffer more in terms of future job progression than employees taking shorter leaves.[62] British women who maintain their employment continuity have been found to be compensated as well as childless women, but women who break their employment after childbirth earn less.[63] Moreover, the longer the leave, the less likely that the employee will ever return to the workplace. In Germany, women who take the full three years' leave (and more, if there is more than one child), often find it difficult to reenter the labor market, and many do not do so.[64] Thus, if the leave is intended to foster continued attachment to the labor force, there is a point of not only diminishing but actually negative returns.

Women's reluctance to be separated from their infants suggests that despite the beneficial workplace impact of shorter leaves, such leaves may impose a higher psychic cost on women. Short maternal leaves have been found, for example, to be a risk factor for depression.[65] As we have seen before, it is hardly surprising that mammalian mothers would feel some psychic distress as a consequence of separation from their infants. Other data show, however, that the strength of a mother's attachment to her infant increases over the first few months after birth,[66] which may suggest that if the mother's goal is to return to work while her child is an infant, perhaps she should do it sooner rather than later.

As to compensation, at least from the short-term perspective of the employee, the more the better. However, while most European countries grant longer leave than currently available under federal law, in most countries the leave is either unpaid or carries a stipend akin to unemployment benefits. Only Sweden, Norway, and Finland maintain a substantial portion of earnings. If the purpose of parental leave is to allow parents to take time with their children without paying the ultimate sanction of having no job when they desire to return to the labor force, then the guarantee of a job upon return may be more important than payment during the leave, although to many families, of course, compensation will be of central concern.

Another important question relates to the source of funding. Although

European systems generally involve some state financing, much of the discussion in the United States has focused on employer-financed leaves. Indeed, to the extent that FMLA leaves are financed through continuation of insurance premiums, employers currently pay for them.

Public financing has several advantages. First, it requires the government to calculate the cost and make the case that the cost is worthwhile. In contrast, employer financing creates an illusion that it is free, when in fact it results in a "hidden tax" that is passed on in the form of lower wages and reduced employment to employees and higher prices to consumers. Second, small employers who have to pay "phantom" employees may be hard-pressed to do so. If support of families is a societal obligation, there seems no reason to require some employers to suffer just because they were unlucky enough to hire several women (or men) who subsequently took extended paid leave. Third, and related to the second, is that employer financing creates a greater incentive, especially for small employers, to discriminate against women of childbearing age in hiring. A small employer, struggling to meet a payroll and faced with two applicants—a woman of childbearing age and anyone else, male or female—can hardly help but take into account that if it hires the woman of childbearing age it may end up having to pay *twice* for that position. Finally, the government is in a better position to recoup costs from women who take leave and then do not return to employment, if a promise to return to work is a condition of receiving the leave.

Fathers and Parental Leave. Parental leave is thought by some to be the key to involving fathers more with their children. Law professor Martin Malin argues, for example, that paid leave is necessary because in most households fathers provide the majority of income and therefore cannot take advantage of the unpaid FMLA leave. He also contends that workplace hostility toward men taking parental leave deters them from taking advantage of leave that is available today. Although Malin provides only limited evidence that there is a large body of men wanting to take parental leave who are deterred by their employers' hostility to it—indeed, the evidence he cites indicates that substantial majorities take some leave following the birth of their children—the possibility that employers are systematically interfering with the exercise of employees' statutory rights may be worthy of further study.

Malin's contention that men are discouraged from taking leaves by the fact that the leaves are unpaid is quite plausible. Compensation, after all, is what causes most people to work in the first place. Men tend to use accrued vacation and personal leave to take time off after the birth, rather than the full twelve weeks (unpaid) provided under the FMLA.[67] If longer leaves were compensated, many men would undoubtedly take them. However, the evidence from Sweden indicates that men would still use them relatively lightly. In Sweden, substantially fewer fathers than mothers take parental leave. Although about three-quarters of Swedish men take some leave, their leaves tend to be much shorter than women's, so that in 1997, 90 percent of paid parental leave days were taken by women. Even with the support given working parents in Sweden, studies find that no more than 10 percent of men shortened their hours after childbirth, in comparison with around 60 percent of women.[68] Ever since adoption of parental leave, Swedish policymakers have been dissatisfied with the level of fathers' participation.[69] Although the extensive paid parental leave available to fathers in Sweden is far beyond anything that has a realistic likelihood of adoption in the United States, even in Sweden the effects in terms of "equal parenthood" have been, according to Linda Haas, "only modest."[70]

Attitudes of both fathers and mothers are important contributors to Swedish fathers' relatively light use of parental leave. Most of the leave available under Swedish law is awarded to the couple, rather than to the individual, and only one parent can take leave at a time. Both mothers and fathers acknowledge that mothers' strong desire to stay home plays an important role in fathers' low use of leave. Linda Haas notes that "mothers have trouble giving up traditional authority for child care, feel guilty about relinquishing child care to their partners, believe men cannot manage alone, and fear a loss of self-respect if they do not make motherhood their primary role." Only one-third of mothers strongly agreed that men should take parental leave. Haas concluded that "men will make little headway in participating equally in parental leave as long as their partners are not really that interested in sharing."[71]

Among Swedish parents who do take leave, Haas reported that fathers are much more likely to report their pleasure at being able to return to work.[72] For women, on the other hand, their return to work is more "bittersweet," because they enjoy their leaves so much. Women were almost twice as likely (63 percent versus 33 percent) to express strong agreement with the statement that parental leave was a welcome respite from their jobs. Interestingly, women with nontraditional sex-role attitudes were actually more likely than women with traditional attitudes to indicate a desire to take parental leave again.

Many Swedish women acknowledge sex differences in attitudes toward leave. The Swedish physician whose comments describing her reduced work involvement after birth were quoted at the outset of this chapter had shared parental leave relatively equally with her husband. Nonetheless, she noted the difference between herself and her husband: "I now prefer a specialty more combinable with family life. I will have to act like I am more interested than I am so I won't get into any trouble. For my husband, career and family are equally important. He doesn't think you have to choose or that there are inherent conflicts."[73]

Because it appears that fathers will not voluntarily take leave at a level deemed desirable, some would like paternal leaves to be made mandatory. Law professor Michael Selmi argues for amending the FMLA to require six-week paid leaves upon the birth of children. In order to increase workplace equality, he argues, it is necessary to "persuad[e] men to behave more like women, rather than trying to induce women to behave more like men."[74] Recognizing that such an authoritarian response is unlikely to draw substantial political support, he advocates as a fall-back position tying federal contracts to employers' success in inducing employees to take parental leave. He does not acknowledge, however, that such an approach is no less authoritarian than a direct mandate. Given the number of employers whose economic health is dependent upon governmental contracts, a contracting-based approach simply means that employees would be forced by their employers (who would be responding to governmental pressures) rather than directly by the government to take leaves.

Unlike Malin, who argues (albeit on less than overwhelming evidence) that increased paternal leaves would be beneficial for fathers and children as well as for women, Selmi's justification rests solely on the equalizing effect that such leaves would have on women in the workplace. He states that "the underlying premise of the idea is that currently women are effectively required to take leave, and this proposal simply would balance that socially imposed condition with one that was legally imposed." However, to the extent that women are obliged to take some leave after birth, it would be more accurate to say that it is a biologically, rather than socially, imposed condition. Beyond a short recuperative leave after childbirth, there is no requirement, socially imposed or otherwise, that a woman take leave after birth, and, as Selmi himself points out, many women return to work soon after giving birth.

Selmi's stated desire is for legislative solutions that "treat women as the norm in the workplace." "Equalizing choices between men and women," he declares in Orwellian fashion, "may entail changing the choices currently available to both men and women." Interestingly, he never actually articulates *why* women should be treated as the norm, although he implies that it is because inequality will persist unless they are. Selmi's requirement that men take six weeks off after the birth of their children just to make them "even" with women, who will take such leave even without a mandate, is reminiscent of Kurt Vonnegut's story *Harrison Bergeron,* in which the government, in the name of equality, required beautiful people to wear ugly masks and smart people to wear radio receivers in their ears that would periodically sound off to disrupt their thoughts.

We seem to have reached a point in civic discourse where it is seldom deemed appropriate to speak of women's obligations in the same way we talk about men's. We have grown accustomed to calls for greater male participation in domestic activities: in the name of equality, husbands ought to do more housework and ought to spend more time with their children even if it means devoting less energy to their careers. Few would think it appropriate to suggest explicitly that in the name of equality, women ought to work more, although that may be the message that many women hear from segments of the feminist movement that have made them feel guilty—as traitors to their sex—for staying home with their children.[75] Even that message is not that homemakers owe it to their families to work more, but rather that they owe it to their sex. A woman, it seems, should be allowed to allocate her attentions to work or family as she likes, and, under this view, we should value her choice irrespective of how she chooses. A parallel is found in discussions of women in the military. The common refrain is that women should be allowed to serve in combat if they want to, rather than that military women should be required to serve in combat under the same terms as military men are obligated to.[76] We have thus gone from a situation in which women were perceived as having all obligations and no choices—when essentially the only role open was wife and mother—to a situation where we view women, unlike men, as having no obligations but wide-open choices.

The fact that this discussion has focused purely on employment issues should not be taken to suggest that only employment issues are important in setting policy. Another issue is the welfare of children. While the impact of day care on the development of infants is a controversial subject, it is a subject implicated in the parental-leave debate. To date, partisans on these questions seem to have adopted somewhat conflicted positions. Many of those—predominantly conservative—who believe that day care is bad for infants and toddlers oppose parental-leave requirements, while many of those—predominantly liberal—who insist that day care is just fine also argue that we need to adopt a parental-leave system "for the children."

Effects of Generous Social Welfare Programs

The United States is often compared unfavorably with Europe in its family policies, but most European schemes were pro-natalist in design, adopted to reduce the level of infant mortality and to stem the tide of population decline rather than to advance the position of women in the workplace.[77] Given the impact that childbearing tends to have on workplace involvement, one might start with skepticism that policies designed to promote childbearing would simultaneously advance the position of women in the workplace, and the European experience tends to validate that doubt.

Sweden's generosity has been a two-edged sword for women's employment. By two measures often viewed as reflective of equity, Sweden seems to do quite well. The female-to-male earnings ratio is far closer to parity than in the United States (0.87 versus 0.72),[78] and Sweden has about the highest female labor-force participation rate in Europe (70 percent versus 60 percent in the United States).[79] There is another side to this story, however. Sweden has the highest level of occupational segregation in Europe, largely due to the creation of so many public-sector "caring" jobs that women overwhelmingly occupy. Moreover, the high level of female labor-force participation masks an important difference between the United States and Sweden: only about 5 percent of employed Swedish women work over forty hours per week, which is a common pattern in the United States,[80] and almost half of all women work less than thirty-five hours per week in accordance with the preference of Swedish women for part-time work.[81] While Swedish women have made a substantial impact on politics—over 40 percent of the Swedish parliament is female[82]—only about 8 percent of managers in Sweden are female, and women are virtually absent from top management. On the other hand, Sweden has been quite successful in increasing its birthrate.[83]

Economists have convincingly demonstrated that mandated benefits are not the "free lunch" that some think.[84] A study of mandated parental leave in European countries found that the economic effects varied with the length of leaves. Where short leaves were mandated, the effect on women's wages was neutral and the effects on women's employment was positive. Where long leaves were available, however, there were negative effects on both wages and employment.[85] Similarly, mandated health-insurance coverage for pregnancy in the United States, which was estimated to cost employers approximately one thousand dollars per employee, resulted in a decline in both female wages and employment.[86]

The economies of Europe seem to have paid a relatively high price for some of these benefits. For example, government spending in Sweden as a share of gross domestic product is approximately twice that in the United States (two-thirds versus one-third).[87] High taxes and employment costs tend to stifle new-business creation and create disincentives to hiring new employees. European unemployment rates have been substantially higher than those of the United States since the early 1980s, and during the late 1990s they were two to three times as high, although many factors other than family policies contributed to that statistic.

While some of the adverse effects are felt by European populations as a whole, others are felt disproportionately by women. Many Swedish women who would prefer not to work do not have that luxury, because high taxes make it difficult to sustain a family on one income.[88] Also, the higher costs associated with female employment create substantial incentives for employers to decline to hire and promote women, and there is some evidence to suggest that employers in high-benefit countries have responded to these incentives.[89]

Family policies have not only economic consequences but social ones as well. The economic function of the Swedish family has waned substantially, for example, as functions previously performed by the family have increasingly been taken over by the state.[90] Not coincidentally, a majority of babies born in Sweden today are born to single mothers, a trend found troublesome by many.[91]

Those who advocate Europe's approach to social welfare as a way to advance women in the work force should consider that women's position in the U.S. work force is typically considered at least a decade ahead of that in Europe. Long parental leaves and readily available part-time work may be desirable on a variety of grounds, but they do not pave the way to the executive suite and high earnings. Indeed, government policies such as child tax credits, family allowances, and paid maternity leave that "subsidize" child rearing have been found to have positive effects on fertility,[92] which implies an attenuation of workplace attachment for women.

Dialogue should continue about whether some of the policies described in this chapter should be pursued. The fact that policies will have little effect—or even negative effects—on statistical measures of women's workplace equality is no reason to reject them, as they may further other policy goals. It would be a mistake, however, to think that policies intended to make childbearing easier are likely to have positive effects on women's employment.

V Sex and the Workplace
Sexuality and Sexual Harassment

The discussion has so far focused predominantly on the effects of individual differences in temperament, ability, and interests on sex differences in workplace outcomes. Beyond sex differences in such attributes, however, is the irreducible fact of sex differences in sexuality, differences that affect the way people interact. Men and women often pursue different kinds of mating strategies (to the extent that they are pursuing mating strategies at all). As a consequence of these different strategies, men and women sometimes have different interests and different perceptions of sexual behavior. Sometimes the results are innocuous enough, but at other times they can result in genuine conflict. When conflict arises in the workplace because of sex or when it takes a sexual form, it is often labeled "sexual harassment," an umbrella term that subsumes a number of behaviors that have distinctly different forms and motivations.

A biological approach to sexual harassment can provide a number of insights lacking in the conventional sociological account. It can assist in predicting the circumstances in which it is most likely to occur, assist in formulation of relevant legal doctrine, and, last but not least, dispel some of the more absurd ideological contentions.

13 Sexual Harassment

The term "sexual harassment" covers a broad array of conduct, and three different kinds of cases will be discussed in this chapter. The first consists of claims that the workplace environment is sexually hostile, based upon the pervasive presence of such things as pin-up pictures, sexual joking, and sexual propositions. The legal standard inquires whether the environment is objectively hostile, which entails an analysis of the reasonableness of the perception of hostility. A recurrent issue raised in the sexual harassment literature and case law is whether reasonableness is evaluated from the perspective of the "reasonable person" or the "reasonable woman." For reasons rooted in evolutionary history, men are likely to perceive women's friendliness as an expression of sexual interest, and women are likely to perceive men's behavior as threatening. An evolutionary analysis therefore reveals the emptiness of the concept of the reasonable person when it comes to sexual matters.

The second category of cases involves sexual overtures from co-workers or supervisors. Much of the social-science literature denies the sexual nature of this conduct, under the oft-repeated slogan that "sexual harassment is about power, not about sex." Given the close relationship over time between sex and power, it is naive to think that they can be so neatly cabined. Under the sociological view, men use sex instrumentally in order to obtain and retain power over women. An evolutionary perspective suggests, however, that the converse is much more nearly correct: men use power instrumentally to obtain sex. Much sexual harassment must be seen as a product of human sexual psychology, a psychology that is itself a product of millions of years of evolution, rather than as simply a product of Western capitalism or an ideology of patriarchy.

The final category, which sometimes overlaps the prior two, involves cases where abusive conduct, whether or not of a sexual nature, is aimed at the victim perhaps because of hostility to the victim or hostility to the victim's sex. Some of the conduct reflects a general tendency of men to attempt to achieve dominance over others. Even prior to entry of women into the work force, men employed such conduct against each other, and they continue to do so in all-male workplaces. The entry of women into previously all-male workplaces has exposed women to preexisting forms of abuse, which may be tailored to their sex but may not be directed at them because of their sex.

One of the inevitable results of sexual integration of the work force has been an expansion of opportunities for sexual interaction and, as a result, sexual conflict. The same conflicting reproductive interests of individual men and women that lead to tensions outside the workplace often lead to even greater tensions within the workplace. Much of this sexual conflict plays itself out as "sexual harassment."

What Is Sexual Harassment?

"Sexual harassment" is a label attached to such a broad variety of conduct that the label has substantially diminished in meaning. Among the actions that courts have said may constitute sexual harassment are such disparate actions as forcible rape; extorting sex for job benefits; sexual touching; widespread sexual jokes; sexually suggestive pictures or cartoons; sexist comments; sexual propositions; romantic overtures; "well-intended compliments"; failure to squelch false rumors that an employee was having an affair with her supervisor; violent attacks on a former lover; attempts to maintain a soured relationship; and harassing actions of a nonsexual form that are directed toward a woman because of her sex.[1] Although these behaviors are widely varied in both form and motivation, they have in common that they are thought to harm either individual women or women as a class.

The breadth and nebulousness of the concept of sexual harassment is due in large part to its origins. Title VII of the Civil Rights Act of 1964, the primary federal law prohibiting sex discrimination in employment, does not expressly prohibit sexual harassment. Instead, it makes it unlawful for an employer "to discriminate against any individual . . . because of such individual's . . . sex."[2] Thus, it has been necessary to characterize the broad variety of "harassing" conduct as sex discrimination in order for the law to reach it. In many cases, such characterization is relatively straightforward; in others, it is highly strained.[3]

Courts have recognized two broad categories of sexual harassment. The first is *quid pro quo harassment,* which typically involves a claim that an employee was required to submit to sexual advances as a condition of receiving job benefits or that failure to submit to such advances resulted in a tangible job detriment. The demand "sleep with me or you're fired" creates an archetypal quid pro quo case. The other form of harassment is *hostile-environment harassment,* which involves a claim that the work environment is permeated with sexuality or "discriminatory intimidation, ridicule, and insult."[4] The plaintiff making a hostile-environment claim must show that she was subjected to "unwelcome" conduct, based upon her sex, that was "sufficiently severe or pervasive to alter the conditions of the victim's employment and create an abusive working environment."

The diversity of conduct that can constitute sexual harassment makes it impossible to develop a unitary view of its causes and therefore of its cures. The absence of a consensus definition is also responsible for the dramatically varying estimates of its prevalence. While some researchers estimate that up to 90 percent of all women have faced some form of sexual harassment in the workplace,[5] surveys also show that most men and women do not think that it is a problem in their own workplaces.[6] The fact that the incidence of harassment declines as its severity increases renders meaningless such broad statements as "approximately 50 percent of female students have been harassed in some way by their professors or instructors, ranging from insulting remarks, come-ons, propositions, bribes, and threats to outright sexual assault."[7] A category of conduct that ranges all the way from an insulting remark to forcible rape may not be a terribly useful one.

Acceptance of the view that various forms of sexual harassment can be understood in evolutionary terms does not, of course, excuse the conduct. Understanding its origins can have benefits, however. It can aid in formulating prevention strategies, which are more likely to be successful if based upon an accurate understanding of offenders' motivations. It can also lead to greater conceptual clarity in the law, which is extraordinarily muddled today, either by leading to judicial clarification of legal doctrine or by suggesting legislative modification of the law tailored more precisely to the perceived evils.

The "Reasonable Woman" versus the "Reasonable Person"

One of the major issues in sexual harassment law concerns the appropriate perspective by which to judge whether the work environment is a hostile one. The Supreme Court requires not only that the plaintiff in fact experience the environment as hostile but also that the perception be objectively reasonable.[8] The question, then, in pursuing this inquiry is whether the reasonableness should be analyzed from the perspective of the "reasonable person" or that of the "reasonable woman." Courts and commentators are divided over the issue.[9]

Choosing between the Standards

The argument for a reasonable-person standard seems to arise out of fear that a reasonable-woman standard is paternalistic and imposes an obligation on men to conform to a standard of conduct that they cannot understand.[10] For example, the Michigan Supreme Court rejected the reasonable-woman standard because "it merely reinforces, and perhaps originates from, the stereotypic notion that first justified subordinating women in the workplace."[11] Such paternalism, said the court, "degrades women and is repugnant to the very ideals of equality that the act is intended to protect."

In contrast, courts adopting the reasonable-woman standard have relied upon just the differences that the Michigan court was reluctant to reinforce. For example, the United States Court of Appeals for the Ninth Circuit reasoned that because "[c]onduct that many men consider unobjectionable may offend many women," a reasonable-person standard "tends to be male-biased and tends to systematically ignore the experiences of women."[12] The court emphasized that because of the focus on the victim, conduct may constitute unlawful sexual harassment even in the absence of any intent to harass on the part of the perpetrator. Indeed, even "well-intentioned compliments" by co-workers or supervisors can form the basis of a sexual harassment cause of action.

We thus have two diametrically opposed points of view. Under one, a sex-conscious standard is said to violate equality principles; under the other, a sex-conscious standard is said to be required by the same principles. This issue is important, of course, only if there are in fact sex differences in perceptions. The literature tends to confirm that there are. Surveys find that men and women differ substantially in their views about some kinds of harassment. Although there is little difference between the sexes in their views of the most serious forms of harassment, they often differ significantly about lesser and more ambiguous forms.[13] While large majorities of both men and women view sexual activity as a requirement of the job to be harassment, larger sex differences appear when it comes to activities such as sexual touching or sexual comments. Sociologist Barbara Gutek found, for example, that while 84 percent of women viewed sexual touching as sexual harassment, only 59 percent of men did so. Similarly, 33 percent of women thought that sexual comments at work that were meant to be complimentary were harassment, while only 22 percent of men thought so.[14] One widely reported finding is that a substantial majority of women would be offended by sexual overtures at work, while a substantial majority of men would be flattered.[15]

Sexual Miscommunication

Because men see the world "through sexual glasses," they tend to see situations as more sexually oriented than women do. A line of psychological studies by Antonia Abbey has shown, for example, that men tend to perceive sexual interest where women perceive only friendly interest.[16] As a result, there is much room for misunderstanding in individual encounters. A woman who has no interest in a sexual relationship with a man may first act in a friendly fashion; the man may interpret the woman's behavior as a "come on" and respond with what he believes are mild indications of sexual interest; the woman may interpret the man's behavior as being friendly and respond with more friendliness; the man may think that she has responded positively to his indications of sexual interest and respond by making a sexual advance that was not desired by the recipient.

Aspects of a woman's appearance may also lead to miscommunication. For example, heavy use of cosmetics is often taken, by both men and women, as a sign of lower levels of morality.[17] Heavy makeup may therefore be interpreted as signaling receptivity to sexual advances, a desirable trait in potential short-term mates.

The differences in perception that lead to miscommunication are easily understood from an evolutionary perspective. As David Buss has explained, "when in doubt, men seem to infer sexual interest, . . . and if over evolutionary history even a tiny fraction of these 'misperceptions' led to sex, then men would have evolved lower thresholds for inferring women's sexual interest."[18] A man who waits to make advances until he is absolutely certain that the woman is sexually interested is not likely to be as reproductively successful as a man who tries as long as there is "some chance." This is especially true given the negative consequences to a woman of being too blatant about her sexual interest.[19]

The risk of miscommunication is enhanced by the perception of many men that women often are just "playing hard to get" and often mean "yes" even if they say "no." Although this notion is often referred to as a "myth,"[20] some women do, in fact, engage in such behavior. A study by Charlene Muehlenhard and Marcia McCoy presented college women with the following scenario:

> You were with a guy you'd never had sexual intercourse with before. He wanted to engage in sexual intercourse and you wanted to also, but for some reason you indicated that you didn't want to, although you had every intention to and were willing to engage in sexual intercourse. In other words, you indicated "no" and you meant "yes."

Although the prevailing ideology maintains that no woman would engage in such behavior, Muehlenhard and McCoy found that a staggering 37 percent of college women responded positively to the question.[21] In 36 percent of these cases, sexual intercourse occurred. In the words of psychologist Linda Mealey, the fact that "females are selected to be coy will mean that sometimes 'no' really does mean 'try a little harder.'"[22]

What might be viewed as the converse of men's bias toward overestimating sexual interest on the part of a woman is women's bias toward overestimating sexual threat on the part of a man in circumstances where there is a risk of coercion. Because of the substantial fitness costs to a woman in our ancestral environment who lost control over both her choice of sexual partner and the timing of reproduction, natural selection would have favored a woman's cautiousness about unwanted sexual attention.[23] Discomfort should begin well before an overt attempt is made, since by that time it may be too late. As a result, women may perceive a man's behavior as threatening if opportunities for escape are limited, even if the man does not intend to convey that message. For very fundamental reasons, then, where a man might see "opportunity," a woman sees "danger."

That men and women differ in their attitudes toward sexual coercion is illustrated by a study of college men's reactions to hypothetical sexual coercion by a woman. Subjects were asked how they believed they would respond to sexual advances ranging in coerciveness from "gentle touch" to being threatened with a knife, by a female initiator who was either "very attractive" or "very unattractive." Subjects reported that they would be more pleased by a knife threat by an attractive woman than a gentle touch by an unattractive woman. The researchers reported surprise that a substantial number of men "viewed an advance by a good-looking woman who threatened harm or held a knife as a positive sexual opportunity."[24] It may be, of course, that the men's responses to real-life situations would be different. In the hypothetical setting, the men may have focused on the sexual dimensions of the situation more than they would have were they actually in physical danger, but quite clearly the men's responses were very different from responses that women would have provided.

A Biological Perspective on the "Reasonable Person"

If a biological perspective can contribute anything to sexual-harassment policy, it must be the insight that a reasonable-person standard is meaningless. When it comes to matters of sex and sexuality, there are no "reasonable persons," only "reasonable men" and "reasonable women." The different sexual natures of men and women cannot be blended into a one-size-fits-all "human sexual nature"; sex must be specified in order to make the concept of reasonableness intelligible.

Although it makes little sense to construct a "reasonable-androgyne" standard, the reasonable-woman standard can itself raise serious problems. In *Ellison v. Brady,*[25] for example, the United States Court of Appeals for the Ninth Circuit held that even though it was quite understandable that the district court had viewed the harasser's conduct as "isolated and trivial" and even though there was no evidence that the harasser harbored ill will toward the plaintiff, the plaintiff's case had merit because the plaintiff's negative reaction was not "idiosyncratic or hypersensitive." The problem with this approach is that even if most reasonable women would not respond the way Ellison did, the court's analysis privileges her view of the matter unless it is so extreme as to be beyond the pale of reason.

The *Ellison* approach allows what may be a small minority of women to set workplace standards for all workers. As one law professor has argued, "even if significant numbers of women enjoy an atmosphere in which sexual jokes abound, if systematically more women than men find this costly, then the practice is discriminatory."[26] Barbara Gutek has found that most people do not consider as harassment "comments or whistles intended to be compliments, quasi-sexual touching such as hugging or an arm around the shoulder, requests for a date or sexual activity often in a joking manner, and sexual jokes or comments that are not directed to a particular person."[27] Yet, many feminists—at least those who write about sexual harassment—take it as self-evident that these behaviors constitute harassment. If 20 percent of men view these behaviors as harassment compared to 30 percent of women, then it is fair simultaneously to say that substantially more women than men see harassment and that most men and most women do not see it. The conduct would constitute harassment under the *Ellison* view, however, because it is difficult to characterize the response of almost a third of all women as "idiosyncratic" or "hypersensitive." Because attitudes toward harassment are strongly correlated with identification with feminist political philosophy,[28] the reasonable-woman standard thus becomes the "reasonable feminist."

Will the Standard Make a Difference?

The debate over the appropriate standard rests implicitly on an assumption that may not be correct: that the substantive legal standard actually makes a real-world difference in courtrooms. Despite the theoretical appeal of viewing the victim's perspective through sex-specific lenses, it is possible that the standard selected will have no impact. Supporting this view is a laboratory study that attempted to measure the effects of selecting between the reasonable-person and reasonable-woman standard.[29] Male and female subjects were presented with scenarios drawn from real harassment cases and asked to evaluate them using the different standards. The judgments of neither males nor females were affected by whether they were applying a reasonable-person or reasonable-woman standard. This may be, as the researchers suggested, that the men lacked sufficient information concerning the "reasonable woman's" perceptions about harassment and therefore relied upon their own attitudes. It seems more likely, however, that the jurors were already incorporating their judgments about the perceptions of women into their decision making under both sets of instruction, since both men and women view coercion of women by men as being more serious than coercion of men by women.[30]

More empirical research could usefully be employed to discern whether the standard does in fact make a difference. If it turns out that judges and juries do not vary their judgments based upon varying formulations of the legal standard, the raging debate over the "reasonable-woman" standard may turn out to be of little more than symbolic importance.

Exclusive Focus on the Victim May Be Inappropriate

Sexual-harassment doctrine may have taken a wrong turn when it began to focus on the perspective of the victim to the exclusion of the perspective of the alleged harasser. In other discrimination cases, the intent to discriminate is central, but many courts view intent (or the lack of it) as wholly irrelevant in harassment cases. Especially if the sexes have different views of what is reasonable, this exclusive focus on the victim may not be appropriate. Moreover, the inherent subjectivity of the standard carries with it a substantial danger of opportunistic charges.[31]

When sex differences in perspective lead to miscommunication, who, if anyone, is to blame? The usual answer in the sexual-harassment literature is that the man is responsible, as he has made an "unwelcome" sexual advance, irrespective of his mental state. One law professor has argued that "our assumption should be [that] sexually oriented conduct or discussions in the workplace are generally demeaning to women and, therefore, improper; [thus]

if sexually oriented conduct occurs, the employer should assume all the risk,"[32] while another has advocated a standard that "rejects the notion that men are entitled to the protection of their misimpressions about how such behavior is interpreted by women.[33] However, a woman's speech, dress, and behavior send important signals to others and affect the way they respond to her. Sometimes these signals are intended; sometimes they are not. As Antonia Abbey has observed, "The behaviors that are misperceived have an ambiguous meaning—smiling may be used to convey friendliness or sexual attraction; revealing dress may be worn simply to look nice or to convey sexual availability."[34]

The ambiguity of cues employed in courtship guarantees that miscommunication will occur frequently, and features of the workplace—such as the need for continued association—especially encourage ambiguity.[35] Because the benefits of this ambiguity are reaped by both men and women, suggestions that explicit verbal consent should precede sexual activity[36]—as in the infamous Antioch College policy—are not realistic. Unless a presumption against sexual interest (or against men) is adopted, there seems little reason to lay the blame exclusively at the feet of the man.[37] Depending upon the circumstances, either the man or the woman may properly bear the lion's share of responsibility, or they may share it equally. Nonetheless, many dismiss with disdain any suggestion that women share with men the responsibility of avoiding this miscommunication. Thus, one law professor is critical of a court's suggestion that a supervisor "must be sensitive to signals from the woman that his comments are unwelcome, and the woman . . . must take responsibility for making those signals clear,"[38] while another expresses scorn for a university official who suggested that women should avoid engaging in "ambiguous conduct."[39]

The focus on the plaintiff's perspective systematically privileges the woman's view, even if the two parties are equally responsible for miscommunication. If men sometimes engage in sexually oriented conduct or speech that a reasonable woman might perceive as threatening even though no threat was intended, the problem could be characterized as either a misperception on the part of the woman or insensitivity on the part of the man concerning the effect of the signals he is sending. By the same token, if women sometimes engage in conduct that a reasonable man might perceive to be inviting even though no invitation was intended, the problem could be characterized as either a misperception on the part of the man or an insensitivity on the part of the woman concerning the signals *she* is sending. In both cases, however, the usual analysis deems the miscommunication the fault of the man, without explanation of why, if men are held responsible for threats they did not intend, women should not be responsible for invitations they did not intend.

If women are to be accepted by their male colleagues on terms of equality, a more restrained approach might be more fruitful. There is some incongruity in saying, on the one hand, that women should serve as combat soldiers, police officers, firefighters, and coal miners, and, on the other hand, that women cannot participate in the labor force on an even footing with men unless the workplace environment is paternalistically modified so that it is acceptable to the most (nonpathologically) sensitive woman. Men often have disputes with coworkers, but, carrying forward a pattern from childhood, men who complain to authority figures rather than working out the problems on their own are often held in contempt by their peers.[40] Women are more likely than men to seek the assistance of supervisors,[41] carrying forward a pattern from childhood in which girls are more likely to seek the protection of parents or other adults, but, by doing so, they risk losing men's respect as equals.

Most people would probably agree that women should be able to seek protection from egregious forms of harassment. However, the law as currently interpreted has pressured employers into adopting "zero tolerance" policies for any matters to which a woman might take sexual offense.[42] The rampant censorship by employers of their employees' speech leads to both resentment on the part of male employees and reinforcement of stereotypes of women as childlike in their need for protection by authority figures.[43] Moreover, the potential of generous recoveries for offense can only increase the amount of offense suffered.

Homosexual Harassment

One class of cases in which a sex-conscious victim-specific standard would likely have a substantial impact involves harassment claims based on an environment that is asserted to be hostile because of homosexual conduct directed toward heterosexuals. In one such case, male employees alleged that their supervisor's distribution of homosexual pornography subjected them to a hostile environment. Although the court concluded that a reasonable person could find the environment offensive, it ruled in favor of the employer because plaintiffs had not shown that the environment was "discriminatorily hostile." If the men objected, it was not because they were men but because they were heterosexuals who "may not be entirely at ease with sexuality in general and homosexuality in particular."[44] The men's "homophobia" is not protected by Title VII, said the court, because the statute's prohibition of sex discrimination does not extend to discrimination on the basis of sexual orientation.

Under the precedents established in cases previously described, the plaintiffs should have been entitled to prevail if they could show that men, to a greater extent than women, are uncomfortable with depictions of male homosexuality or that the supervisor showed them the pictures because they were men. Certainly, the court would not have rejected a hostile-environment claim by a woman on the ground that her objection to pornography in the workplace demonstrated that she was "not entirely at ease with sexuality." Most men, at least in our society, are uncomfortable around homosexually oriented speech and behavior, and it seems that the discomfort of men exceeds that of women.[45]

The ultimate cause of male unease with homosexuality may be similar to the origins of women's greater discomfort around heterosexually oriented speech and behavior. A major reason for women's greater objection to sexually related themes is concern about maintaining control over their own reproductive decisions, loss of which can result in a substantial decrease in reproductive fitness. Consequently, women are inherently wary about conduct that may result in a loss of that choice. Men are not vulnerable in the same way. Because of their greater strength, men are less likely to be forced by women into sexual encounters; because women are operating under different incentives, they are less likely to try to force men into sexual encounters; and because of differences in reproductive biology, a momentary loss of sexual control by men does not present the same risks faced by women. Men face no danger of pregnancy from homosexually oriented conduct, of course, but they may face another threat to their reproductive fitness. Men's reproductive success is, as we have seen, substantially affected by their status in male hierarchies. A man who is subjected to homosexual advances, at least in American culture, stands to lose status by being perceived as sexually attractive to members of his own sex or as being homosexual himself. Being especially sensitive to that kind of status impairment, men may respond negatively and forcefully to homosexual overtures—even if they generally do not have antihomosexual attitudes[46]—despite the fact that men rarely respond violently to unwanted heterosexual overtures.

The law has recognized the strength of many men's adverse reactions to homosexual advances in other contexts. For example, under the criminal law, a defendant in a murder action may sometimes raise a provocation defense where the victim made a homosexual advance to the defendant. If accepted by the jury, the defense reduces the crime from murder to manslaughter. The rationale for the defense, which is what the law calls a "partial excuse" rather than a justification, is that homosexual advances may cause the "ordinary" man—though not the "reasonable" one—to lose self-control.[47]

A highly publicized example of such a reaction occurred when Jonathan Schmitz fired two shotgun blasts into the chest of Scott Amedure, who three days earlier had revealed during a taping of *The Jenny Jones Show* that he had a homosexual "crush" on Schmitz. According to Schmitz's father, Jonathan was concerned that everyone would think he was a homosexual. After the shooting, Jonathan called 911 to turn himself in, telling the operator that he had just shot a man. When asked why, he replied in near hysterics, "He f——ked me on national TV."[48]

There is a striking parallel between the Schmitz case and the trivial altercation homicides studied by Martin Daly and Margo Wilson and discussed in chapter 9. These homicides arise out of disputes that observers would view as inconsequential, such as a minor jostling or petty insult. Like the participants in Daly and Wilson's trivial altercations, Schmitz killed Amedure in front of a witness. Because it would have made no sense from Schmitz's perspective to kill Amedure unless he was publicly associated with the killing, he immediately turned himself in. While Schmitz's reaction was extreme—his father, perhaps understatedly, characterized him as having a "thin skin"—the fact of a negative reaction under such circumstances is not at all unusual, and the public nature of the forum no doubt escalated the strength of the reaction.

Because of sex differences in responses to homosexuality, a sexual-harassment case involving homosexual advances toward a man should be analyzed under a "reasonable-man" standard. The trier of fact should ask not how equivalent advances by a woman would be received by a man, nor should it ask how a woman would respond to equivalent advances by a man. Rather, the trier of fact should ask how the reasonable man would respond to sexual advances from another man. Not surprisingly, research indicates that men report more negative reactions to unwanted sexual contact by males than by females.[49] While some might feel that it is "heterosexist" to assume that the "reasonable man" is heterosexual or is averse to homosexual advances, that is not a normative assumption but rather a descriptive statement: the overwhelming majority of men are heterosexual and react negatively to homosexual advances. One can similarly assume that the "reasonable woman" is not a "nymphomaniac" without making value judgments about nymphomania.

Power versus Sex

It seems to be an article of faith with many that sexual harassment is not "about sex" at all, but "about power." Thus, we hear that "the goal of sexual harassment is not sexual pleasure but gaining power over another,"[50] that "sexual harassment has always been primarily about power, only rarely about sex, and never about romance,"[51] and that "sexual harassment is about the abuse of power. It is not about sex."[52] Why it has to be one or the other is not clear, but many writers go out of their way to create the impression that harassment victims are not selected according to criteria of sexual attractiveness but rather chosen at random for the sufficient reason that they are women—victims of a male need to oppress women. For example, Barbara Gutek asserts that harassment "is likely to happen to almost any female worker"; it can happen to the "young and old, rich and poor, professional and unskilled."[53] Another group of researchers notes that "some individuals believe only attractive women are sexually harassed," but that "empirical studies do not support this belief, since women in all ranges of attractiveness have reported harassment."[54] Notwithstanding the care with which this impression is fostered, Gutek herself acknowledges that most victims are young and either single or divorced.[55] In hostile-environment cases, of course, the plaintiffs may be of any age or description because the harassing conduct may not be aimed at anyone in particular; pinup pictures, for example, potentially create a hostile environment for all female employees.

The power-versus-sex argument implies that power and sex are mutually exclusive explanations. However, even in the clear quid pro quo case—the kind of case that most conspicuously involves power—intuitively it seems to be about both: a supervisor is using his workplace power to extort sex. To say that it is only about power makes no more sense than saying that bank robbery is only about guns, not about money.

Sexual Harassment as a Mating Strategy

A biological perspective should lead to skepticism about a claim that activities that result in sexual intercourse are not "about sex" and especially so when the claim is that power and sex are unrelated. Throughout human history, men have used power as a way of obtaining sex, whether coercively or through making themselves more attractive as mates; indeed, sexual coercion is widespread throughout the animal kingdom.[56] Men with the most power in history—despots whose subjects lived at their sufferance—almost invariably surrounded themselves with large numbers of nubile women whose favors they could command at their pleasure.[57] Male "despots" in the workplace often adopt a similar strategy, as revealed in the ample cases involving men who are alleged to have used their workplace power to coerce sexual favors.[58] In the first sexual-harassment case to reach the Supreme Court, for example, the plaintiff alleged that she had had unwelcome sexual intercourse with her supervisor forty to fifty times over the four-year period of her employment.[59]

What is the evidence for the surprising, yet common, suggestion that a man's extorting sex from a woman is not "about sex?" The answer is, "not much." The research most commonly invoked to support the conclusion is a study of sexual harassment performed by Sandra Tangri and her colleagues. They proposed and tested three potentially explanatory models of sexual harassment: the "natural/biological" model, which views harassment as a consequence of natural physical attraction; the "organizational model," which views harassment as a consequence of hierarchy within the organization; and the "sociocultural" model, which views sexual harassment as a result of sex-role socialization and the differential distribution of power in the larger society. They concluded that there was less evidence to support the natural/biological model than there was to support the other two models.[60] (Needless to say, the three explanations are not mutually exclusive.) Following the Tangri study, the idea that there is any significant biological contribution to harassment is usually mentioned just to be dismissed.[61] One law professor, for example, has asserted (without citation of any authority) that biology "has absolutely nothing to do with" sexual exploitation of women by male supervisors.[62]

Profiles of Victims and Harassers. Rejection of the natural/biological model followed from a failure of the observed data to satisfy the predictions that the researchers derived from the model. They had predicted that if the model were correct, harassers and victims would be of both sexes; victims would be similar to their harassers in age, race, and occupational status; both harasser and victim would be unmarried; and the victim would be the only person to whom the harasser directs his attention.[63] They also predicted that the behaviors would resemble courtship behaviors, they would stop once the victim indicated a lack of interest, and victims would be "flattered" by the behaviors. Because their data did not satisfy those expectations, they rejected the model.

Rejection of the natural/biological model was inappropriate, because the researchers' predictions were not actually derivable from the model. The study was largely designed to reject the model if the data suggested that harassers were not looking for *wives*, but no one has suggested that sexual harassment is mostly "about marriage." What the researchers should have tested was whether victims of harassment tend to possess those traits that would cause them to be viewed as either long-term *or* short-term mates.

Had Tangri and her associates asked the right questions, they probably would have been less eager to reject the natural/biological model. Psychologists Michael Studd and Urs Gattiker subsequently analyzed the sexual harassment literature using a more sophisticated view of sexual psychology.[64] Their predictions concerning the profiles of victims and harassers are largely those that would be predicted from Buss and Schmitt's study of sexual strategies, discussed in chapter 2.

The predicted profiles vary to some extent depending upon whether one expects the actor to be employing long-term or short-term strategies. However, in either event, the strongest prediction is that the harasser is male and the victim is female. Men are usually the sexual initiators in both kinds of relationships, and men more often than women resort to short-term sexual strategies. Probably the next strongest prediction, regardless of which kind of strategy is employed, is that the preferred victim is of reproductive age, and one might also expect her to be relatively attractive. She is also likely be unmarried, because that may indicate to the potential harasser both that she is "available" and that she does not have the protection of another male. Therefore, one would expect the ideal victim to be single, divorced, widowed, or, if married, married to a man whom the potential harasser believes not to be a threat.

The above predictions are largely satisfied, with the most predictive variable for both harasser and victim being sex. Although estimates of the frequency of sexual harassment vary substantially, all agree that the overwhelming proportion of victims are women and the overwhelming proportion of harassers are men. According to the EEOC, over 86 percent of sexual harassment charges in 2000 were filed by women.[65] Unfortunately, the EEOC's statistics do not disclose the sex of the alleged harasser, but many claims by men involve male harassers, either homosexual or heterosexual. A review of the case law strongly suggests that the number of men complaining of harassment by other men substantially outstrips the number of men complaining of harassment by women (and the number of women complaining of harassment by other women is extremely small), so the fraction of charges filed by men would substantially exceed the number of charges involving female harassers.

The predictions derived from evolutionary reasoning hold true for victim attributes other than sex as well. For example, most victims are under the age of thirty-five, and they are disproportionately single, divorced, or separated.[66] From these profiles, Studd and Gattiker concluded that the motivation of most male harassers is sexual, although not romantic: "Male harassers appear to be motivated more by the continuing possibility of coercing women into sexual intercourse than by the desire to build a long-term relationship built on mutual and congruent emotional interest."[67] They also note, however, that some of these behaviors may start out as "legitimate" sexual or romantic attraction that later degenerates into harassment.

Vulnerability. Those committed to rejection of a sexual explanation for sexual harassment argue that the reason that men select unattached women of reproductive age is that they are more vulnerable rather than because they are sexually attractive. For example, Tangri and associates concluded that "the tendency of individuals with greater degrees of personal vulnerability and dependence on their job to experience more harassment [is some of] the strongest evidence available in these data against the natural model."[68] Under this view, young unattached women are particularly vulnerable, and it is simply coincidental that these also happen to be women who would be sexually attractive to a potential harasser. However, it is not clear why a finding that victims tend to be vulnerable would undermine the natural/biological model. To the extent that the man's strategy is to convert his workplace power into satisfaction of his sexual urges, one would expect him to focus on victims susceptible to the exercise of that power. It is not just attractiveness that is important to him; it is attractiveness plus accessibility.

Vulnerability of the victim may be a necessary condition for harassment, but it is not a sufficient one. Although the harassment literature often assumes that younger women are vulnerable because they lack the economic power or maturity of judgment to resist advances of a supervisor, when the discussion turns to age discrimination it is generally assumed that older workers are more vulnerable and dependent upon their jobs than young people.[69] As between a twenty-five-year-old single waitress with six months of seniority and a fifty-five-year-old widowed secretary who has put in thirty years on the job and who has no pension, who is in the better position to walk off the job and get a new one that is just as good? Most people would say the twenty-five-year-old. Which of them has the greater likelihood of experiencing harassment? All the data suggest it is also the waitress.

If "vulnerability" is the determinative factor, one would expect a surfeit of older victims. Indeed, if there is truth to the common assertion that racial minorities, the disabled, and homosexuals are especially vulnerable to employer oppression, one could construct a detailed profile of the ideal victim under a vulnerability analysis: she would be an elderly, handicapped, black lesbian. Needless to say, the data are to the contrary.

Cases Involving Female Harassers. If sexual harassment is just about organizational power and not about sexual psychologies, it is odd that women so rarely sexually coerce their subordinates. Some argue that it is because women seldom possess the necessary power.[70] However, one need not be a top executive to possess the power to employ sexual coercion; all one needs is a higher position in the hierarchy than the target. In fact, most litigated cases involve relatively low-level supervisors. Millions of women have the organizational power to extort sex if they want to. While it might be suggested that, given the readiness of men to engage in casual sex, women do not need to coerce them, that response itself rests on the different sexual psychologies of men and women. There is also little evidence that women supervisors engage in frequent sexual relations with their subordinates, and Buss and Schmitt's analysis of sexual strategies would suggest that this would not be common, given women's preference for higher-status men.

Sexual harassment claims brought by women differ in important ways from those brought by men, not only in quantity but also in quality. The facts in cases with a female plaintiff and a male accused harasser tend to resemble short-term strategies, while those in cases with a male plaintiff and a female accused harasser tend to look more like long-term strategies. Quid pro quo claims brought by female plaintiffs are of several varieties:

1. the plaintiff refused to engage in sexual relations with the harasser and was thereafter subjected to retaliation;
2. the plaintiff engaged in sexual relations with the harasser after being threatened with adverse consequences if she refused; or
3. the alleged harasser was seeking to initiate a consensual relationship or prolong one that had turned sour.[71]

A review of hundreds of such cases leaves the firm impression that cases in the first two categories are dramatically more common than those in the third category and that a desire for "raw sex" is a major motivation for much of the harassment.

The pattern is quite different in those relatively few harassment cases brought by male plaintiffs against female supervisors, with the cases much more likely to fall into the third category. There is almost always some hint of either a prior consensual sexual relationship or a desire on the part of the female harasser for a relationship that is more than purely sexual. For example, one case involved an allegation that the female harasser had reacted spitefully when the consensual romance between the plaintiff and her had ended;[72] another involved a claim by a man who had broken off a relationship with his supervisor with whom he had been living and with whom he had a daughter;[73] and another involved a man who was fired after rejecting the overtures of his female supervisor who had grabbed his buttocks, made sexually suggestive comments to him, and told him that she loved him and wanted him to move in with her so she could stay home and take care of his son.[74]

Additional Sociocultural Explanations

The view that extorted sex is purely a consequence of a hierarchical power structure and completely unconnected to sexual psychology is simply untenable. The weakness of that argument has given rise to other arguments that have in common that they continue to discount the importance of sexuality.

Society-wide Sex-based Power Differentials. The next tactic of those unwilling to let go of the power explanation is to retreat to the "sociocultural model" and argue that sexual harassment of women by men is not so much a product of power differences between individual men and women in organizations, but rather of power differences between men and women in society at large.[75] Yet the sociocultural model does not explain homosexual harassment. If heterosexual harassment is really about men as a class dominating women as a class, harassment of men by homosexual male supervisors must be a completely different phenomenon. Homosexual supervisors who impose quid pro quo demands on male subordinates are not oppressing women, and they are not oppressing men as a class in any meaningful sense. Instead, they are using their workplace power to satisfy their sexual desires in the same way that a heterosexual supervisor might.

One variant of the sociocultural theory holds that sexual harassment is an attempt by men to exert power because of their fear that women constitute a threat to men's economic or social standing.[76] However, this theory does not explain the most pervasive coercive sex in our history in the master-servant relationship: sexual relations between a slave owner and his slaves. Female slaves constituted no threat to their owner's economic or social standing; indeed, quite the opposite was true. Moreover, slave owners had no need to prove who the "boss" was, and, even if they did, they had a wide range of tools at their disposal. Nonetheless, sexual relations between slave and owner were extremely common.[77] One Louisiana planter expressed his belief that there was not "a likely-looking black girl in this state that is not the paramour of a white man."[78] Indeed, the pervasiveness of such relationships was one of the principal objections of many abolitionists to the institution of slavery.[79]

Slave owners did not seek slave women at random for sexual relations. Rather, they preferred those having the attributes that men typically value in sexual partners: reproductive value as demonstrated by youth and beauty. In her narrative, ex-slave Harriet Jacobs noted that "[i]f God has bestowed beauty upon her, it will prove her greatest curse. That which commands admiration in the white woman only hastens the degradation of the female slave."[80] Similarly, Wilma King observed that "[t]he more comely a slave girl, the greater the possibility that she would experience sexual abuse and sale as a 'fancy girl' for illicit purposes."[81] This greater desirability was reflected in price. Historian Eugene Genovese reports that in New Orleans a prime field hand would sell for $1,800, a first-class blacksmith would bring $2,500, while a "particularly beautiful girl or young woman might bring $5,000."[82]

The view that men as a class are combining to oppress women as a class is a pervasive flaw in much of the literature. In her book *Sexual Shakedown,* for example, Lin Farley described sexual harassment as a means for men to "insure their domination of modern work, hence society, because the patriarchy cannot lose control of its material base."[83] Another pair of researchers, addressing the finding that harassment victims are more likely to be single or divorced, suggested that while the rule that "women became the property of their husbands upon marriage, and as such, were off-limits to men . . . may no longer be overtly expressed, there is evidence that an implicit understanding of this principle still exists among men."[84] Barbara Gutek similarly asserts that "[p]erhaps women who are married or widowed are viewed as being under a man's protection that other men honor"; other women, she says, are "fair game."[85]

The notion that there is some implied agreement among men to avoid trespassing onto the "turf" of other men is fanciful. An attractive married woman interested in extramarital relations will find no shortage of men ready and willing to trespass onto her husband's "property," as evidenced by the large numbers of women who have extramarital affairs.[86] The suggestion that male harassers, the majority of whom are married, are more inclined to honor the marital vows of other men than they are to honor their own, is odd, to say the least. This reasoning also fails to account for Beth Schneider's finding that "closed" lesbians—who might have a male partner for all the harasser knows—are subjected to more sexual advances than "open" lesbians—whose partners are known to be women—unless the argument is that men view the open lesbians' partners as "honorary men."

The view that there is a male conspiracy against women neglects the fact that harassment of women harms men also. The women who are victims of harassment are all daughters of some man, and many are wives, sisters, and mothers of others. Still other women are involved in long-term relationships with men. Whose side are these men on in this supposed male-female conflict? Perhaps some of these men are harassers themselves, but that simply illustrates that sometimes the interests of members of the two sexes coincide, and sometimes they conflict.[87]

"Sex-Role Spillover." Barbara Gutek has proposed an additional explanation for sexual harassment, what she calls "sex-role spillover," which is the "carryover of gender-based expectations into the workplace."[88] Being a "sex object" is one of the characteristics of femaleness, she says, and "women are assumed to be sexual and to elicit sexual overtures from men rather naturally." The problem with the theory of "sex-role spillover" is the problem that plagues social-role theory generally. It assumes that roles are simply assumed by individuals, and it implies that they can be switched on and off at will. Gutek's model implies that workers should shed their "sex role" when they get to work and assume their "work role." She argues that "if people at work behaved within the narrow confines of work roles [instead of pursuant to sex roles,] then sexual jokes, flirtatious behavior, sexual overtures, and sexual coercion would not exist in most workplaces."[89] But male workers are still male, and female workers are still female; sex is more than just a "social characteristic," as Gutek labels it.[90] Workers do not become asexual robots upon crossing the workplace threshold; they bring their psyches and their hormones with them. The fact that men act like men and women act like women in the workplace is, as computer programmers like to say, "a feature not a bug." It is not a reflection of a flaw in the contemporary workplace; it is an inevitable biological fact of life. This is not to say that employees should behave at work as if they were in a singles' bar, but it does mean that the completely desexualized workplace favored by some is unlikely to be achieved.

Is "Exploitation" a Two-Way Street?

The picture of the male supervisor abusing his power by making sexual advances to a subordinate is not the complete picture. Just as men use their power and resources to obtain sex, women may use sex (or the promise of it) to obtain power and other resources, even in the workplace. Robert Quinn, in his study of workplace romances, noted a common pattern that he called the "utilitarian relationship": one in which "the male is perceived as seeking such things as excitement, ego satisfaction, adventure, and sexual experience, while the female is viewed as in search of organizational rewards."[91] Barbara Gutek also observes that some women engage in "flirtatious and seductive behavior" in order to "get in good with the boss,"[92] and women, especially in blue-collar environments, sometimes use flirtatious behavior to avoid difficult work.[93]

One group of researchers has pointed out the implicit sexism in the traditional view of sexual harassment: the assumption "that females are helpless pawns and males are omnipotent despots in relationships."[94] They note, for example, that over one-third of female college students acknowledged having flirted with their male professors and conclude that "courtship or quasi-courtship between powerful older males and younger females is not solely due to the operation of male choice, but is also driven by female choice."[95] In a sense, men and women are quite similar: both use what they have to get what they want. Men often use their economic and physical power to get sex from women, and women often use their sexual power over men to gain economic or other benefits. The exploitation of women by men gets much attention, but it should be noted that exploitation can be a two-way street. The vulnerability of men to sexual exploitation is well captured in the Yiddish proverb "Ven der putz shteht, ligt der sechel in drerd," which, translated loosely, means "When the penis is erect, the brains get buried in the ground."[96]

Organizational prohibitions on romantic or sexual relationships between those of different rank in the hierarchy are often advocated as a means of protecting women against sexual harassment by their superiors. In some situations, they can in fact serve that function. Such rules do more, however, than protect women from men. They also protect high-status women in organizations from sexual competition from lower-status (and often younger) women.[97]

A More Nuanced View of the Relationship between Power and Sex

The relationship between power and sexual harassment is considerably more subtle than is often appreciated. To be sure, there are numerous cases in the literature where the connection between power and sexual advances is quite explicit. When a supervisor expressly links a woman's job benefits to her submission, both parties understand the nature of the transaction. In other cases, however, the relationship is more nuanced. Indeed, it has been estimated that 75 percent of harassers do not realize that they are harassing.[98]

John Bargh and Paula Raymond have suggested that many men in supervisory positions do not realize they are exploiting their power because for them there is an unconscious link between power and sex.[99] Subjects scoring high on measures of likelihood of engaging in sexual harassment or other sexual aggression respond faster to sex words when they are subliminally primed with a power word. Such men also rate a woman's attractiveness higher if they have first been primed with a power stimulus. Men who are low on measures of sexual aggression show no such priming effect. Although Bargh and Raymond did not attempt to discern the origin of the association, they attribute it to learning; they say that the association comes about because power has become "habitually associated" with sex. It is likely, however, that the two have become "associated" more primally in the psyche of many men without any extended learning. As psychologist James Dabbs has observed, "The major social effect of testosterone is to orient us toward issues of sex and power."[100]

According to Bargh and Raymond, when a man who has this association is put in a position of power over a woman, the association becomes activated, and it tends to bias both his interpretation of the woman's behavior and his perception of her attractiveness. The man will not be aware that it was his power that triggered his thinking in terms of sex; he will assume that his interest was triggered by the usual factors of the woman's attractiveness and behavior. In other words, his power over her may act upon him in the same way that signs of sexual accessibility do. Henry Kissinger's famous remark that "power is the great aphrodisiac" is often taken to mean that a man's power is sexually arousing to women, but it may be a source of sexual arousal for the man holding it as well.

The finding that many men have an automatic association of power and sex suggests that modification of sexual harassment training may be in order. Much of that training is focused on warning men not to exploit their power over subordinates to coerce sex or, more generally, that sexual relationships between supervisors and subordinates are inappropriate. Neither of these messages is likely to be terribly effective in modifying the behavior of a man having the power-sex association. Because he is not aware that his power has created the attraction, he will not view his conduct as exploitative. Moreover, because the man may perceive the relationship as one of mutual attraction, he will not view it as wrong. This is not to suggest that such messages will have no effect at all, as men are more likely to engage in sexual harassment in circumstances in which such behavior is accepted or condoned.[101] Perhaps a more effective strategy, however, is to educate men specifically about the association of power and sex. Teaching men that being in a position of power will sometimes result in erroneous perceptions may tend to counter their bias. At least to some extent, forewarned may be forearmed.

Power is unquestionably an important component of some kinds of sexual harassment. Power is an essential ingredient of quid pro quo harassment, since the harasser must have the power to carry through on his threat if he is rejected. But power is only part of the explanation. The focus on power to the exclusion of sex appears to be an unfortunate side effect of the fact that most of the scholarship on harassment has been from the woman's, if not the feminist's, point of view. From the perspective of the victim, it may seem like all power and no sex. But it is the perspective of the harasser that must be understood in order to fashion an effective response, since it is his motivations and perceptions that control his actions, and it is his actions that the law seeks to regulate.

Ultimately, whether sexual harassment is "about power" or "about sex" may turn out not to matter in shaping policies; that is, the same policies may follow whatever the outcome of this debate. However, we are more likely to settle on an effective response if we have a full explanation of the cause of the behavior, an explanation that involves the intersection of power and male sexual psychology. It is about power, but it is even more fundamentally about sex.

Not All "Sexual Harassment" Is Sex Discrimination

At the risk of apparent inconsistency, the discussion now moves from arguing that much sexual harassment that feminists have viewed as being about power is really about sex to arguing that much harassment that feminists think is about sex is really about dominance—which may have little to do with the sex of the target—or about hostility that may not necessarily be based on sex.

The kind of sexual harassment described in the preceding section involved men whose goal was sexual relations. The men either consummated the relationship or gave other indications that this was their purpose, such as by retaliating against the women who rejected them. In those cases, one need not invoke any explanation other than the men's desire for sexual relations and a willingness to use their power to achieve their goal. Other cases typically included unquestioningly under the rubric of sexual harassment involve expressions of hostility toward a woman that take a sexual form. The New Jersey Supreme Court, for example, has declared that "when the harassing conduct is sexual or sexist in nature," the "because of sex" element will be automatically satisfied.[102]

Men's tendency toward competition and striving for positions in hierarchies often leads to behaviors that, especially to women, may seem quite harsh. As Joan Kennedy Taylor has observed, "Men will harass, tease, and verbally abuse each other, find vulnerable spots and use them to fluster each other—almost automatically. When called on it, they will say it was all in fun. Women, when faced with such behavior, tend to take the content seriously, rather than identifying the underlying game."[103] Women may therefore perceive a hostility that was not intended and interpret that perceived hostility to be based upon their sex.

Many sexual harassment cases do involve outright hostility, but that hostility does not necessarily flow from sex-based animus. Men may call women vulgar sexual names and make crude overtures that on their face merely look like sexual advances, if the latter term is defined to mean speech or conduct entered into in the hope that sexual relations will result. When a man points to his genitals and says to a woman, "Here, sit on this,"[104] or points to hers and says, "Give me some of that stuff,"[105] it is clear that he has no expectation that his behavior will entice the woman into engaging in sexual activities with him. These statements may reflect animus toward women as a class, perhaps animated by their entry into formerly all-male workplaces, but they may not.[106] That is, they may be equivalent to a man's saying to another man, "Come on over here and I'll kick your ass," which, on its face, appears to be an invitation but in reality may simply be an assertion of dominance. Yet courts seldom feel it necessary to consider whether any animus expressed is class based, as would seem to be required under a discrimination theory. For example, in one case the plaintiff complained that sexually explicit pictures were posted with her name on them, someone wrote "the wicked witch is gone" on the calendar when she took vacation, someone tried to set her up on a false drug-buying charge, and she was repeatedly told that women did not belong on the police force.[107] Although the hostility toward the plaintiff may have been motivated by her sex, the court's findings indicated that the leading harasser knew her before she started work, and when she began work he indicated that he hated her. The court found in her favor without addressing the question whether the harassment was motivated by sex-based or rather merely personal animus.

Employers have frequently lost cases despite their defense that harassing conduct was not sex discrimination because the harassing supervisor was abusive to all employees. For example, in *Steiner v. Showboat Operating Co.*, the employer argued, in defense against the plaintiff's assertion that her supervisor had used highly offensive language toward her, that the supervisor was also abusive to men.[108] The court rejected that argument, reasoning that although the supervisor was abusive to men and women alike, his abusive treatment and remarks to women "were of a sexual or gender-specific nature." The court stated, "While [the supervisor] may have referred to men as 'assholes,' he referred to women as 'dumb fucking broads' and 'fucking cunts.' Thus, his abuse of men in no way related to their gender, [but] his abuse of female employees, especially [the plaintiff], centered on the fact that they were females."

The court may have been correct that the supervisor in question was especially hostile to the plaintiff because of her sex, but that conclusion does not follow automatically from the presence of sex-based language. Sex is a highly salient characteristic, and when it comes to vulgar insults, our language is quite sex-specific. One need only think of the vulgar epithets that flow so freely in our culture to appreciate that very few of them are characteristically applied equally to both sexes. One study found a very high degree of sex specificity in insults.[109] Male and female subjects were asked to name the worst thing that a man or woman could be called. Among the eight most frequently named insults, there was no overlap between the target sexes. The terms "bastard," "faggot," "gay," and "prick" were identified for male targets, and the terms "bitch," "cunt," "slut," and "whore" were identified for female targets. In *Steiner,* the court found that harassment was sex based in part because the manager called men "assholes," which the court viewed as "in no way related to their gender," while his abuse of female employees "centered on the fact that they were females." While anatomically the court might be correct that the insult directed toward the men was not sex-specific, it is, nonetheless, an epithet reserved largely for men. It is an odd legal standard that makes the question of whether Ms. Steiner was subjected to sexual harassment turn on whether the supervisor called her male co-workers "assholes" instead of "pricks."

Interactions in the workplace can sometimes be unpleasant. Many people, particularly men, are prone to overtly aggressive behavior toward those they dislike or perceive to be vulnerable. Whether or not their dislike is based upon sex-based animus, their cruel behavior may have sexual overtones, in large part because of the sexualized worldview that men tend to possess. It should be remembered that men's quest for dominance has not been primarily about attaining dominance over women, but achieving dominance over other men, which is consistent with Barbara Gutek's finding that in the workplace "women are less often treated disrespectfully than men are."[110] Women may be particularly disadvantaged in these kinds of conflicts—which typically do not involve lone males attacking other lone males, but rather groups of males, the membership of which sometimes changes, that prey on the weak—because men seem more inclined to create these opportunistic coalitions than women.[111]

A great deal of what women perceive (or claim to perceive) to be harassment based on their sex may in fact be ritual hazing that all employees in some workplaces undergo, which is not necessarily based on hostility at all. Senior employees often use hazing to establish their dominance and to push new employees into "letting go of their old identities with and loyalties to former groups and organizations and taking on new identities."[112]

Hazing is usually aimed at perceived vulnerabilities. If a woman is known to be easily upset by sexual talk, the hazing is likely to take that form. Hazing has traditionally been a mostly male experience. Because women tend to have less experience with it, they are more likely to believe that they are being singled out and to view it as sex-based harassment. Although both men and women are subject to hazing, the greater the perceived difference between the newcomer and the group (whether sex, race, background, age, education, or religion), the more testing of the newcomer there is likely to be. In mostly male environments, therefore, women tend either to undergo more hazing than men or "they are protected from hazing and thereby are not given a chance to prove themselves and thus gain true membership."[113]

Even when women are subjected to the same hazing as men, they often respond to it differently. Women are more likely than men to become visibly angry or upset, a reaction that often elicits more hazing of both males and females. Moreover, women are more likely to seek assistance from their supervisors, which is often a poor strategy because it is perceived as disloyal to the group and may create a permanent alienation from it.[114]

Psychologist Louise Fitzgerald complains that "virtually millions of women are subjected to experiences ranging from insults to assault—many on an ongoing or recurrent basis—as the price of earning a living."[115] While this is a true statement, it is no less true if the word "women" is replaced with the word "men." This long-standing phenomenon was not viewed as a social problem, however, until women began to experience what men have always experienced. A comprehensive understanding of human psychology and behavior reveals that much of what is commonly assumed to be sex discrimination may be something else altogether—men treating women the way they have long treated each other—an outcome that many feminists claim to have wanted all along.

Conclusion

The concept of sexual harassment has become a Procrustean bed into which a number of disparate phenomena have been forced, only some of which can fairly be called sex discrimination. Invoking the concept of sexual harassment, many feminist writers have urged a "desexualization" of the workplace, although it is not clear that most women actually desire this result. In fact, because of the proximity of men and women in the workplace, organizations are "a natural environment for romantic relationships."[116]

A realistic view of human nature suggests that as long as men and women inhabit the same workplaces, they will interact as human beings. Part of the way that human beings interact is sexually and romantically. Referring to the notion that a clear split between work life and private life exists, sociologist Beth Schneider has written:

> Although this set of ideas still permeates sociological and everyday thought, it is not necessarily consistent with either individuals' opinions or the reality of their daily work or sexual lives. For generations, sociologists have found that work relationships are far from simply task-oriented, but the source of friendship and informal social ties with consequences for workers as well as their workplaces.[117]

Although it is easy enough to criticize workplace romances, it is not clear that it would be a better world if more men and women met at bars rather than in the workplace.

Much writing about sexual harassment ignores the twin facts that women are active participants in their sexual lives rather than simple pawns of men and that the workplace can legitimately be a locus of sexual attraction. Although sexual harassment surveys typically ask whether respondents have ever been subjected to unwelcome sexual advances in the workplace, they seldom ask whether they have been subjected to welcome ones. The answer would likely often be in the affirmative, since large numbers of workers find their romantic partners at work.[118]

The *tabula rasa* perspective of human nature—the view that sex is just a "social construct"—has led many to believe that people can simply be educated to leave their sexual psychologies behind them: "Check your sex role at the door, and pick up your work role and your time card inside!" Yet sexual interactions (and sexual conflict) are inevitable consequences of the propinquity of reproductive-age males and females. As Lionel Tiger has illustrated, modern organizations—the military being an extreme example—increasingly create environments that exacerbate this conflict.[119] So, at the same time that forces are mobilizing to combat sex discrimination, many of these same forces are combining to increase sexual harassment.

An easy response is to "get tough" on sexual harassment, as the increasingly frequent "zero tolerance" policies are meant to do. Such policies, however, have substantial costs for both men and women. Liability standards have forced employers to impose rigid speech codes on employees in violation of the First Amendment[120] and have led to reduced morale on the part of employees who feel they are walking on eggshells when they deal with members of the opposite sex. Fear of accusations of harassment discourage informal contacts—such as drinks after work—that have often benefited subordinates in their relationships with their superiors. Moreover, vigorous prosecution of sexual harassment laws creates an additional incentive for employers to decline to hire women because potential liability creates an additional way in which women are more expensive to employ than men.

Given the decline of many of the social norms and institutions designed to protect women from men, it is difficult to be sanguine about an acceptable solution. It is almost certainly the case that as long as this conflict is viewed as primarily ideological, rather than biological, successful resolution—or even optimal management—will continue to be elusive.

14 Conclusion

The evidence and arguments put forward in this book will be troubling to many. Some may believe that invocation of biology is implicitly (or perhaps even explicitly) a defense of the status quo—a paean to the virtue of existing arrangements or at least a testament to their inevitability. The defense, however, is more limited. It is that many of the workplace patterns that are laid at the foot of nefarious causes such as discrimination by employers or sexist socialization have causes that are less invidious and less attributable to an antifemale ideology than is commonly recognized.

A consensus about the causes of workplace patterns does not foreordain consensus about policy responses. One's values are important, and values are not directly derivable from scientific fact. Proponents of laissez-faire policies will likely draw free-market implications, while those more inclined toward governmental intervention may settle on more activist approaches. Everyone interested in workplace policy, however, whatever his political or social outlook, should desire an accurate understanding of the underlying causes of current patterns.

It would be a mistake to interpret average temperamental or cognitive sex differences as limitations on the potential of individual girls and women. Nothing contained in this book implies that women cannot or should not be corporate presidents or theoretical physicists, only that equal representation of women in these positions is unlikely to occur unless selection processes are modified with the specific purpose of guaranteeing proportional representation.

Sufficient overlap exists on most traits that there are few occupations that should be expected to remain the exclusive domain of one sex, but many occupations will remain overwhelmingly male or overwhelmingly female if people continue to select occupations on the basis of their preferences and abilities. Expansion of the choices available to women (and to men) increases the influence of individual preferences on workplace outcomes. To the extent that individuals' preferences differ, we should expect them to seek different workplace rewards. Because the average endowment of men and women differs—in temperament, cognitive ability, values, and interests—it would be astonishing if their occupational preferences and behaviors were identical.

Modern attitudes about preferences are somewhat conflicted. The value that Western liberals place upon individual liberty rests heavily on the assumption that the preferences of individuals differ. Each individual should be free, within broad limits, to pursue his own ends. There is, therefore, something vaguely illiberal about both the assumption that all individuals *should* have the same preferences and attempts to ensure the outcomes that would result if they did.

Some people believe that even if sex differences exist, there is harm in publicizing them because they can become self-fulfilling prophecies. Even if the "correct" ratio of professional mathematicians is, say, 5 males to 1 female, it is harmful to make that fact widely known, because then mathematics will be labeled a "male field," and girls will assume that it is *only* for males. Although that is a rational concern, it is not well supported empirically. Clark McCauley found, for example, that when asked to estimate the proportion of males or females in a number of sex-stereotyped occupations, subjects showed no evidence of stereotypic exaggeration. The correlation between estimates and actual percentages was high, indicating that people rank-ordered them accurately, but where the subjects erred, it was almost always in the direction of underestimating the difference between men and women in the occupation.[1] Similarly, Mary Ann Cejka and Alice Eagly found that participants systematically underestimated the extent to which male-dominated and female-dominated occupations were segregated.[2]

It may seem odd that this book implies the near-inevitability of disproportionate male representation at the highest levels in corporate and other hierarchies, at least under current incentives, at the same time that other writers are predicting seemingly contrary trends. In 1999, two books appeared on the market, coincidentally both by Rutgers University anthropologists: *The First Sex* by Helen Fisher and *The Decline of Males* by Lionel Tiger. Both chronicled changes in the workplace, in education, and in broader social forces such as increasing female control over reproduction. Fisher's book emphasized the positive—the ascendancy of females—but gave little attention to the social effects of the "displaced males" that ascendancy of females implies. Tiger analyzed many of the same trends, but his view was more pessimistic, as the specter of large numbers of marginalized males does not bode well for any society. Neither Fisher's nor Tiger's analysis is inconsistent with that provided here, however. As noted in chapter 6, the gender gap in compensation shrinks with changes in work that favor women. Nonetheless, men will continue to dominate the scarce positions at the top of hierarchies as long as it is necessary to devote decades of intense labor-market activity to obtain them, even if women come to predominate in middle-management positions and even if men also disproportionately occupy the bottom of hierarchies. Men will similarly continue to dominate math-intensive fields, as well as fields that expose workers to substantial physical risks.

The extent of one's willingness to live with the sex differences in outcomes described here depends to some extent on one's definitions of equality. If current workplace outcomes are a cumulative consequence of millions of individual choices made by men and women guided by their sexually dimorphic psyches, are the outcomes of those choices rendered suspect because those sexually dimorphic minds incline men and women to make their choices in systematically different ways? This question resembles, if not entails, the familiar question of whether the equality that ought to be of importance to policy makers is "equality of opportunity" or "equality of result." Those who place primary importance on equality of opportunity may say that as long as both men and women are given the opportunity to pursue the opportunities that the workplace provides, the outcomes are unimportant. Those who look to group outcomes, on the other hand, may say that the critical question is what the different groups end up with. However, we cannot say that the "outcome" for women is deficient without specifying with precision what that outcome is. We cannot, that is, simply look at women's income and occupational attainment without also considering what they get in return for the occupational tradeoffs that they make.

The question of agency is at the core of this book. Are women, like men, active agents in their own lives, making rational decisions based upon their own preferences? Or are they pawns of both men and society—making suboptimal "choices" that are forced on them by others? All indications are that the former is closer to the mark. Women, though somewhat constrained by life circumstances, as are men, make rational and responsible choices that are most compatible with their temperaments, abilities, and desires.

People often ask what "women" (meaning women as a group) think of the ideas developed in this book, the apparent expectation being that women would be universally hostile to them. The answer is, of course, that women as a group are varied and have as broad a range of opinion on these questions as men do. Many women, not surprisingly, do find these ideas offensive, if not threatening. Seemingly just as many, or perhaps even more, however, are quite open and indeed vigorous in their support, as they seem to recognize themselves or others in the descriptions. However, the underlying basis for this understandable question is misguided. The issues discussed in this book are not "women's issues," but "human issues." Every woman is some man's daughter, and most are some man's sister, wife, or mother; every man is some woman's son, and most are some woman's brother, husband, or father. No matter what one's view of the causes or consequences of the current structure of the workplace, we should reject once and for all the notion that these are issues of "men versus women" or "women versus men."

Notes

Chapter 1. Introduction

1. *King Solomon's Mines,* 1985.
2. Hume, 1999, 86–87.
3. Mead, 1935, 279.
4. Bernard, 1945.
5. Mead, 1928.
6. Mead, 1935, 280.
7. Watson, 1925, 74–75.
8. Fausto-Sterling, 1993, 20.
9. Lorber, 1993, 569.
10. Brown, 1991.
11. D'Andrade, 1966.
12. Browne, 2001a.
13. Eagly and Wood, 1999a.
14. Furchtgott-Roth and Stolba, 1999, 21.
15. Browne, 1998a, 1998b, 1995a.
16. Lewontin, Rose, and Kamin, 1984, ix–x.
17. Kay, 1990, 75.
18. Lott, 1996.
19. Browne, 1984.
20. Jones, 1999, 1997.
21. Fausto-Sterling, 1992, 11–12; Kitcher, 1985, 9.

Part I. How the Sexes Differ

Chapter 2. Sex Differences in Temperament

1. Williams and Best, 1990, 77–78.
2. Jussim, McCauley, and Lee, 1995.
3. Fox, 1992.
4. Eagly, 1995, 154.
5. Eagly and Steffen, 1986; Rushton et al., 1986.
6. Archer, 1991; Eagly and Steffen, 1986.
7. Archer, 1991.
8. Ahlgren, 1983; Lynn, 1993.
9. Weinberg and Ragan, 1979.
10. Hoyenga and Hoyenga, 1993, 319.
11. Mirowsky and Ross, 1995.
12. Lundberg, 1983.
13. Ahlgren and Johnson, 1979.
14. Lever, 1976.
15. Hrdy, 1981, 130.
16. Maccoby, 1998, 39.
17. Maccoby, 1998, 39.
18. Hrdy, 1981, 129.
19. Mazur and Booth, 1998.
20. DiPietro, 1981.
21. Charlesworth and La Freniere, 1983.
22. Benenson et al., 1997
23. La Freniere, Strayer, and Gauthier, 1984.
24. Maccoby, 1998, 29.
25. Whiting and Edwards, 1988.

26. Maccoby, 1998, 27.
27. Jacklin and Maccoby, 1978.
28. Omark and Edelman, 1975.
29. Pettit et al., 1990.
30. Benenson, 1999; Savin-Williams, 1979.
31. Wilson and Daly, 1985, 69.
32. Hoyenga and Hoyenga, 1993, 321.
33. Hughes, Sullivan, and Beaird, 1986.
34. Dweck and Bush, 1976.
35. Miller, 1985.
36. Corbin, 1981; Croxton, Chiacchia, and Wagner, 1987.
37. Argote et al., 1976.
38. Arch, 1993.
39. Lundgren, Sampson, and Cahoon, 1998.
40. Colarelli, Alampay, and Canali, 2001.
41. Megargee, 1969.
42. Palmer and Folds-Bennett, 1998.
43. Meara and Day, 1993.
44. Weisfeld, 1986.
45. Byrnes et al., 1999.
46. Marcusson and Oehmisch, 1977.
47. Veevers and Gee, 1986.
48. Ginsburg and Miller, 1982.
49. Morrongiello and Dawber, 1998.
50. Morrongiello and Rennie, 1998.
51. Hillier and Morrongiello, 1998.
52. Horvath and Zuckerman, 1993.
53. Schrader and Wann, 1999.
54. Gregersen and Berg, 1994.
55. *New York Times*, October 3, 1993.
56. Toscano and Windau, 1998.
57. Johnson, 1996.
58. McClelland and Watson, 1973.
59. Begum and Ahmed, 1986.
60. Arch, 1993, 5.
61. Horner, 1972.
62. Atkinson, 1957.
63. Atkinson et al., 1960.
64. Geary, 1998; Thompson and Walker, 1989.
65. Badger, Craft, and Jensen, 1998; Freedman and DeBoer, 1979.
66. Goldberg, Blumberg, and Kriger, 1982.
67. Pleck, 1997.
68. Maccoby, 1998, 261–262; Rossi, 1977.
69. Hewlett, 1988.
70. Amato and Booth, 1996.
71. Hoffman, 1977, 715.
72. Adams, Summers, and Christopherson, 1993.
73. Feingold, 1994; Rushton et al., 1986.
74. Goodenough, 1957.
75. McGuinness and Symonds, 1977.
76. Josephs, Markus, and Tafarodi, 1992.
77. Moir and Jessel, 1989, 157.
78. Gilligan, 1982, 17.
79. Buss and Schmitt, 1993.
80. Baumeister, Catanese, and Vohs, 2001.
81. Laumann et al., 1994, 533–534.
82. Clark and Hatfield, 1989.
83. Hrdy, 1999, 224.
84. Laumann et al., 1994, 162.
85. Laumann et al., 1994.
86. Brewer et al., 2000.
87. Buss and Schmitt, 1993.

Chapter 3. Sex Differences in Cognitive Abilities

1. Halpern, 2000; Kimura, 1999.
2. Kimura, 1999, 43–78.
3. Kimura, 1999, 91–101.
4. Browne, 1993a.
5. Halpern, 2000, 68–73.
6. Feingold, 1995; Jensen, 1998, 535.
7. Willingham and Cole, 1997, 68–69.
8. Lubinski and Benbow, 1992.
9. Jensen, 1998, 537.
10. Jensen, 1998, 536–543.
11. Jensen, 1981, 289.
12. Kimura, 1999, 53–54.
13. Voyer, Voyer, and Bryden, 1995. Also Silverman, Phillips, and Silverman, 1996; Collins and Kimura, 1997.

14. Kimura, 1999, 50–51; McBurney et al., 1997.
15. Silverman and Eals, 1992.
16. Barnfield, 1999.
17. Kimura, 1999, 47–48.
18. Moffat, Hampson, and Hatzipantelis, 1998.
19. Silverman et al., 2000.
20. Malinowski and Gillespie, 2001.
21. Liben, 1995; Zernike, 2000.
22. Dabbs et al., 1998.
23. Wilberg and Lynn, 1999.
24. Stanley and Stumpf, 1996.
25. Kimura, 1999, 68.
26. Jensen, 1998, 535.
27. College Board, 2000.
28. Lubinski and Benbow, 1992, 62.
29. College Board, 2000.
30. Minor and Benbow, 1996.
31. Colangelo and Kerr, 1990.
32. Marshall and Smith, 1987.
33. Low and Over, 1993.
34. Halpern, 2000, 82.
35. Benbow, 1988, 218.
36. College Board, 1997.
37. National Science Foundation, 1999, 27.
38. Lubinski and Benbow, 1992, 62.
39. Carretta, 1997.
40. Hedges and Nowell, 1995.
41. Huttenlocher et al., 1991.
42. Kimura, 1999, 91–92.
43. Willingham and Johnson, 1997, Table S–4.
44. National Center for Education Statistics, 2000, 4.
45. Kimura, 1999, 94–95.
46. College Board, 1997; Halpern, 2000, 128.
47. Backman, 1979.
48. Halpern, 2000, 121.
49. Rosenthal and Rubin, 1982.
50. Feingold, 1988; Halpern, 2000, 127.
51. Masters and Sanders, 1993.
52. Hedges and Nowell, 1995.
53. Albington and Leaf, 1991.
54. Kimura, 1999, 69–70.
55. Halpern, 2000, 124–127.

Part II. Women in the Workplace

Chapter 4. The Glass Ceiling

1. Lyness and Thompson, 1997.
2. Browne, 1998a.
3. Glass Ceiling Commission, 1995a, 31.
4. Townsend, 1996.
5. Glass Ceiling Commission, 1995a, 10.
6. Glass Ceiling Commission, 1995a, 3.
7. Townsend, 1996.
8. Reiser et al., 1993.
9. Morrison et al., 1992, 28–32.
10. Gasse, 1982, 58.
11. Maccoby, 1976.
12. Wong, Kettlewell, and Sproule, 1985.
13. Hennig and Jardim, 1977, 76–93.
14. Schwartz, 1992, 226.
15. Schwartz, 1989, 67.
16. Lord, de Vader, and Alliger, 1986.
17. Nicholson, 1998, 146.
18. Eagly et al., 1994.
19. Lynch and Post, 1996.
20. Zimmer, 1988.
21. Williams, 1992.
22. Grey and Gordon, 1978.
23. MacCrimmon and Wehrung, 1990, 433.
24. Sexton and Bowman-Upton, 1990; Subich et al., 1989.
25. Olsen and Cox, 2001.
26. Arch, 1993, 8.
27. Hennig and Jardim, 1977, 27.
28. Fischer, 1976.
29. Townsend, 1996.
30. Glass Ceiling Commission, 1995a, 13.
31. Hoffmann and Reed, 1982.
32. Harrell, 1993.
33. Schwartz, 1992, 45, 61.
34. Udry and Eckland, 1984.
35. Eccles, 1987, 245.

36. Williams, 1991, 1602–1603.
37. Weisman et al., 1986.
38. Townsend, 1989.
39. Baber and Monaghan, 1988.
40. Fox-Genovese, 1996, 215.
41. Valian, 1998.
42. Goldberg, 1968.
43. Swim et al., 1989.
44. Olian, Schwab, and Haberfeld, 1988.
45. Powell and Butterfield, 1994.
46. Cobb-Clark and Dunlop, 1999.
47. Browne, 2001a.
48. Simmons, 2001.
49. Reis, Young, and Jury, 1999.
50. Fierman, 1990; Moskal, 1997.
51. Morrison et al., 1992, 48–53.
52. Eagly and Johnson, 1990.
53. Davis, 1992, 8.
54. Fisher, 1999.
55. Lever, 1976.
56. Kline and Sell, 1996.
57. Bartlett, Grant, and Miller, 1990.
58. Maccoby, 1998, 40. Also Charlesworth, 1996.
59. Brophy, 1989. Also Buttner and Rosen, 1988.
60. Interagency Committee for Women's Business Enterprise, 1996.
61. Devine, 1994.
62. Devine, 1994.
63. Hundley, 2001; Parasuraman et al., 1996.
64. Glass Ceiling Commission, 1995a, 35.

Chapter 5. Occupational Segregation

1. Littleton, 1987.
2. Goldberg, 1993, 36.
3. England, 1979; Gottfredson, 1981.
4. Farrell, 1993, 105.
5. Krantz, 1992.
6. Lubinski, 2000.
7. Dawis and Lofquist, 1984.
8. Bretz and Judge, 1994.
9. Schwartz, 1989, 67.
10. Holland, 1997.
11. Gottfredson and Holland, 1996; Holland, 1997, 268–271.
12. Donnay and Borgen, 1996; Kaufman and McLean, 1998.
13. Hansen, 2000.
14. Kaufman and McLean, 1998.
15. Aros, Henly, and Curtis, 1998.
16. Lubinski, 2000.
17. Harmon and Borgen, 1995.
18. Kaufman and McLean, 1998.
19. Cronin, 1995; Douce and Hansen, 1988.
20. Hansen, 1988, 139.
21. Lubinski, Benbow, and Sanders, 1993, 701.
22. Holloway, 1993, 95.
23. National Science Foundation, 1999, 262.
24. Brown and Silverman, 1999; National Center for Education Statistics, 2000, 72.
25. Holloway, 1993, 96.
26. Kimura, 1999, 76.
27. National Center for Education Statistics, 2000, 24; National Center for Education Statistics, 2000, 34, 40.
28. National Science Foundation, 1999, 14–16.
29. Collier, Spokane, and Bazler, 1998.
30. Seymour and Hewitt, 1997.
31. National Science Foundation, 1999, 158.
32. National Science Foundation, 2000, 120.
33. National Science Foundation, 1999, 171.
34. Halpern, Haviland, and Killian, 1998.
35. Benbow and Lubinski, 1993.
36. Lubinski and Benbow, 1992.
37. Lubinski et al., 1993.
38. Lubinski and Benbow, 1992
39. Humphreys, Lubinski, and Yao, 1993.
40. Humphreys and Lubinski, 1996.
41. Alper, 1993.

42. Trankina, 1993.
43. Collier, Spokane, and Bazler, 1998.
44. Kirshnit, Ham, and Richards, 1989.
45. Adair, 1991.
46. Strenta et al., 1994.
47. Baker, 1998.
48. Farmer, 1988.
49. National Science Foundation, 1999, 40.
50. Seymour and Hewitt, 1997.
51. Barinaga, 1994.
52. Kahn, 1994.
53. Barinaga, 1994; Vitug, 1994.
54. O'Farrell, 1999.
55. Deaux and Ullman, 1982.
56. O'Farrell, 1999.
57. Padavic, 1992.
58. Padavic and Reskin, 1990.
59. Ragins and Scandura, 1995.
60. Ragins and Scandura, 1995.
61. Balkin, 1988.
62. Browne, 2001a.
63. Padavic and Reskin, 1990.
64. Loscocco, 1990.
65. Holland, 1997, 268–269.
66. Deaux and Ullman, 1982.
67. Pheasant, 1983.
68. Bishop, Cureton, and Collins, 1987; Pheasant, 1983.
69. Lillydahl, 1986.
70. Padavic, 1991.
71. Toscano, 1997.
72. Flynn, 2000.
73. Valliere, 2000.
74. Flynn, 2000.
75. Weinberg, 2000.

Chapter 6. The Gender Gap in Compensation

1. Department of Labor, 2001.
2. O'Neill and Polachek, 1993.
3. *New York Times*, January 31, 1999.
4. Schwartz, 1992, 279.
5. Groshen, 1991.
6. Kachigan, 1991, 181.
7. Polachek, 1987, 66.
8. Paetzold and Willborn, 1994, §6.09.
9. O'Neill and Polachek, 1993.
10. Frank, 1978.
11. Huang, 1997.
12. Mansnerus, 1993; Nossel and Westfall, 1998.
13. Lal, Yoon, and Carlson, 1999.
14. National Science Foundation, 1996.
15. National Science Foundation, 1996, 75.
16. Mealey, 2000, 350–352.
17. Solberg and Laughlin, 1995.
18. O'Neill and Polachek, 1993.
19. Hattiangadi and Habib, 2000, 33.
20. Fuchs, 1988, 47–48.
21. Burke, 1998.
22. Harrell, 1993.
23. Harrell and Alpert, 1989.
24. Markham and McKee, 1995; Scott and McLellan, 1990.
25. Kilbourne and England, 1996.
26. Bartlett and Miller, 1985.
27. Subich et al., 1989.
28. Chauvin and Ash, 1994.
29. Ziller, 1957.
30. Smith, 1776, 111.
31. Filer, 1985.
32. Farrell, 1993, 105.
33. Filer, 1985.
34. Duncan and Holmlund, 1983; Filer, 1985; Hersch, 1991.
35. Tam, 1997.
36. Daymont and Andrisani, 1984; Hecker, 1998.
37. O'Neill, 1990.
38. Shaw and Shapiro, 1987.
39. Eide, 1994; O'Neill, 1990.
40. Glass Ceiling Commission, 1995a, 152.
41. Glass Ceiling Commission, 1995b.
42. Paglin and Rufolo, 1990.
43. Gerhart, 1990.
44. Hamermesh, 1993.
45. Bellas, 1994.
46. Hoffmann and Hoffmann, 1987.
47. Kilbourne and England, 1996, 69.

48. Hellerstein and Neumark, 1998; Hellerstein, Neumark, and Troske, 1996.
49. Goldin, 1990, 104; Rhoads, 1993, 140, 160.
50. Xie and Shauman, 1998; Zuckerman, 1991.
51. Cole and Singer, 1991.
52. Barnett et al., 1998. Also Tesch et al., 1995.
53. Carr et al., 1998.
54. Palepu et al., 1998.
55. Kaplan et al., 1996.
56. Reiser et al., 1993. Also Leibenluft et al., 1993.
57. Long, 1992.
58. Astin, 1991; Zuckerman, 1991.
59. Lindgren and Seltzer, 1996.
60. Cole and Fiorentine, 1991, 191, 223.
61. Daymont and Andrisani, 1984.
62. Long, 1995.
63. Murray and Atkinson, 1981.
64. Daymont and Andrisani, 1984.
65. Martin and Kirkcaldy, 1998.
66. Kirkcaldy and Furnham, 1993.
67. Bridges, 1989; Daymont and Andrisani, 1984; Lacy, Bokemeier, and Shepard, 1983.
68. Frank, 1995.
69. Bridges, 1989; Konrad et al., 2000; Murray and Atkinson, 1981.
70. Hinze, 2000; Redman et al., 1994; Shye, 1991.
71. American Association of Nurse Anesthetists, 2001.
72. Bradley, 1989, 201.
73. Shye, 1991.
74. Shye, 1991.
75. Margolis, Greenwood, and Heilbron, 1983.
76. Weisman et al., 1986.
77. Winter and Butters, 1998.
78. Brown and Silverman, 1999; Phillips-Miller, Morrison, and Campbell, 2001.
79. Betz and O'Connell, 1987.
80. Tanner et al., 1999.
81. Tanner et al., 1999, 113.
82. Reskin and Roos, 1990.
83. American Association of Colleges of Pharmacy, 2001.
84. Betz and O'Connell, 1987.
85. Stuhlmacher and Walters, 1999.
86. Callahan-Levy and Messé, 1979.
87. Major and Konar, 1984; Major, Vanderslice, and McFarlin, 1985.
88. Brenner and Bertsch, 1983.
89. Yanico and Hardin, 1986.
90. Blau and Kahn, 1992.
91. Furchtgott-Roth and Stolba, 1999, 14.
92. Korenman and Neumark, 1994, 1992; Long, 1995.
93. Kim and Polachek, 1994; Mincer and Ofek, 1982.
94. Polachek, 1995.
95. O'Neill and Polachek, 1993.
96. Nakosteen and Zimmer, 1987.
97. Korenman and Neumark, 1991.
98. O'Neill and Polachek, 1993.
99. Fuchs, 1986.
100. McElrath, 1992.
101. Becker, 1985; Hersch, 1991.
102. Coverman, 1983.
103. Baber and Monaghan, 1988.
104. Primack and O'Leary, 1993.
105. Frank, 1978.
106. Schwartz, 1992, 229.
107. Deitch and Sanderson, 1987; McElrath, 1992; Shauman and Xie, 1996.
108. Primack and O'Leary, 1993.
109. Merritt, Reskin, and Fondell, 1993.
110. Glass Ceiling Commission, 1995a, 151.
111. O'Neill and Polachek, 1993.
112. Fuchs, 1988, 14–15.
113. Tolbert and Moen, 1998.
114. Hunter, 1993, 148.
115. Department of Labor, Women's Bureau, 1998.
116. Garton and Pratt, 1991.
117. Gibbons, Lynn, and Stiles, 1997.
118. Kirkcaldy and Cooper, 1992.
119. Roback, 1993, 122.

Part III. Origins of Sex Differences

Chapter 7. Why Socialization Is an Inadequate Explanation

1. Goldsmith, 1991, 7.
2. Schooley et al., 1994.
3. Wells and Hepper, 1999.
4. Fausto-Sterling, 1992, 153.
5. Kohlberg, 1966.
6. Hoyenga and Hoyenga, 1993, 218–219.
7. McGuinness, 1990.
8. Leinbach and Fagot, 1993.
9. Campbell et al., 2000.
10. Goldberg and Lewis, 1969.
11. Eaton and Enns, 1986.
12. Servin, Bohlin, and Berlin, 1999.
13. Koot and Verhulst, 1991.
14. Caldera, Huston, and O'Brien, 1989; Fein et al., 1975; O'Brien and Huston, 1985.
15. Harper and Sanders, 1975.
16. Pedersen and Bell, 1970.
17. Campbell, Muncer, and Odber, 1998.
18. Berenbaum and Hines, 1992.
19. Blakemore, LaRue, and Olejnik, 1979; Maccoby and Jacklin, 1974, 363.
20. Eisenberg, Murray, and Hite, 1982.
21. Guttentag and Bray, 1976, 299.
22. Maccoby, 1987, 235.
23. Jacklin and Baker, 1993; Lytton and Romney, 1991.
24. Also Scarr and McCartney, 1983.
25. Campenni, 1999; Fisher-Thompson, 1993.
26. Maccoby, 1987.
27. Bell and Harper, 1977, 53.
28. Maccoby and Martin, 1983, 60.
29. Hennig and Jardim, 1977, 76–93.
30. Williams, 1989, 11.
31. Benenson, Philippoussis, and Leeb, 1999.
32. Lever, 1976, 485–486.
33. Halpern, 2000, 82; McGuinness, 1979.
34. Farrell, 1993, 15.
35. Burke, 1989.
36. Kleinfeld, 1998; Sommers, 2000, 20–23.
37. Altermatt, Jovanovic, and Perry, 1998.
38. Kohlberg, 1966.
39. Tooby and Cosmides, 1990.
40. Marks, 1987, 229.
41. Tiger, 1979.
42. Mealey, 2000, 192.
43. Harris, 1998.
44. Zuckerman, 1987.
45. Lamb, Easterbrooks, and Holden, 1980.
46. Fagot, 1984.
47. Hoyenga and Hoyenga, 1993, 316.
48. Fagot, 1985; Maccoby, 1990.
49. Bigler and Liben, 1990.
50. Guttentag and Bray, 1976, 299.
51. Matteson, 1991.
52. Sandberg and Meyer-Bahlburg, 1994.
53. Fagot, 1984.
54. Bergen and Williams, 1991.
55. Eagly and Steffen, 1984.
56. Lueptow, Garovich, and Lueptow, 1995.
57. Colapinto, 2000; Diamond and Sigmundson, 1997.
58. Costa and McCrae, 1987.
59. Kerns and Berenbaum, 1991.
60. Gibbs and Wilson, 1999.
61. Johnson and Meade, 1987.
62. Halpern, 1997.
63. Mann et al., 1990; Silverman, Phillips, and Silverman, 1996.
64. Halpern, 2000, 122.
65. National Center for Education Statistics, 1998.
66. Gilger, 1996.
67. Levine et al., 1999.
68. Nordvik and Amponsah, 1998.
69. Hyde, Fennema, and Lamon, 1990; Robinson et al., 1996.
70. Raymond and Benbow, 1986.
71. Baenninger and Newcombe, 1989.
72. Halpern, 2000, 108.
73. Benbow and Stanley, 1996, 259.
74. McGuinness, 1988.

75. Lummis and Stevenson, 1990.
76. Kimball, 1989.
77. Lubinski and Benbow, 1992.
78. Halpern, 1996.
79. Tiger and Shepher, 1975, 262–263.
80. Tiger, 1987.
81. Spiro, 1996, 106.

Chapter 8. Hormones: The Proximate Cause

1. Reinisch, Ziemba-Davis, and Sanders, 1991.
2. Arnold, 1996; Haqq et al., 1994.
3. Mealey, 2000, 14.
4. Breedlove, 1994.
5. Nyborg, 1994, 34.
6. Dittmann et al., 1990b.
7. Fitch and Denenberg, 1998.
8. Breedlove, Cooke, and Jordan, 1994.
9. Zucker et al., 1996.
10. Dittmann et al., 1990a.
11. Reinisch et al., 1991.
12. Berenbaum and Hines, 1992; Berenbaum and Snyder, 1995; Hines and Kaufman, 1994.
13. Berenbaum and Resnick, 1997.
14. Leveroni and Berenbaum, 1998.
15. Berenbaum, 1999.
16. Zucker et al., 1996.
17. Quadagno, Briscoe, and Quadagno, 1977.
18. Berenbaum, 1999.
19. Dittmann et al., 1990b.
20. Berenbaum and Resnick, 1997.
21. Berenbaum, 1999.
22. Masica et al., 1971.
23. Fratianni and Imperato-McGinley, 1994.
24. Nyborg, 1994, 74.
25. Reinisch et al., 1991, 270.
26. Udry, Morris, and Kovenock, 1995.
27. Udry, 2000.
28. Goy, Bercovitch, and McBrair, 1988.
29. Ellis, 1986.
30. Ward et al., 1996.
31. Christensen and Breedlove, 1998.
32. Dabbs, 1990.
33. Nyborg, 1994, 36.
34. Ellis, 1986.
35. Dabbs et al., 1995.
36. Schiavi et al., 1984.
37. Purifoy and Koopmans, 1980.
38. Mazur and Booth, 1998.
39. Flinn et al., 1996; Tremblay et al., 1998.
40. Cohen, 1998.
41. Cashdan, 1998, 1996.
42. Archer, 1991.
43. Allee, Collias, and Lutherman, 1939.
44. Diamond and Young, 1963.
45. Joslyn, 1973.
46. Katz and Konner, 1981.
47. Hampson, Rovet, and Altmann, 1998.
48. Iijima et al., 2001.
49. Grimshaw, Sitarenios, and Finegan, 1995.
50. Gouchie and Kimura, 1991.
51. Moffat and Hampson, 1996.
52. Janowsky, Oviatt, and Orwoll, 1994.
53. Hausman, 1999.
54. Nyborg, 1994, 100–110.
55. Hampson, 1990.
56. Kimura, 1999, 119–120.
57. Hier and Crowley, 1982.
58. Slabbekoorn et al., 1999; Van Goozen et al., 1995, 1994.
59. Miles et al., 1998.
60. Williams and Meck, 1991.
61. Lacreuse et al., 1999.

Chapter 9. Evolutionary Theory and the Ultimate Cause

1. Bonner, 1980; Goldsmith, 1991.
2. Waal, 2001, 69.
3. Darwin, 1859.
4. Clutton-Brock,1985;LeBoeuf,1974.

5. Darwin, 1871, vol. I, 256.
6. Cronin, 1991, 234.
7. Trivers, 1972, 139.
8. Clutton-Brock, 1991.
9. Konner and Worthman, 1980.
10. Buss, 1994, 200–201.
11. Buss and Schmitt, 1993.
12. Buss and Schmitt, 1993, 218–219.
13. Betzig, 1993, 1986.
14. Gregersen, 1982, 84.
15. Daly and Wilson, 1988, 127–128.
16. Mazur, Halpern, and Udry, 1994, 88.
17. Eagly and Wood, 1999a.
18. Berglund, 1993.
19. Hedrick and Dill, 1993.
20. Buss, 1994, 46; Townsend, 1987.
21. Buss et al., 2001.
22. Felmlee, 1994.
23. Low, 1992.
24. Smuts, 1987, 402.
25. Maccoby, 1998, 261–262.
26. Duberman, 1975, 50.
27. Silverman et al., 1993.
28. Miller, 2000, 192–194.
29. Packer et al., 1995.
30. Wood, Corcoran, and Courant, 1993.
31. Low, 1989.
32. Bjorklund and Kipp, 1996.
33. Smuts, 1995, 5.
34. Buss, 2000, 189.
35. Rice, 1996.
36. Geary, 1996. Also Kolakowski and Malina, 1974.
37. Silverman and Eals, 1992.
38. Gaulin and FitzGerald, 1986; Gaulin, FitzGerald, and Wartell, 1990.
39. Geary, 1996.
40. Hawkes, 1991.
41. Ghiselin, 1996.
42. Kimball, 1996.
43. Hammer and Dusek, 1996.
44. Geary, 1998, 312–313.
45. Kimura, 1999, 70. Also Casey et al., 1995.
46. Foss, 1998.
47. Halpern, 2000, 121; Hausman, 1999.
48. Tiger, 1988.

Part IV. Public Policy

Chapter 10. Difference or Disadvantage?

1. Dowd, 1989; MacKinnon, 1990.
2. Mahony, 1995, 5.
3. Mahony, 1995, 216.
4. Kalb and Hugick, 1990; Tanner et al., 1999.
5. Dreher and Ash, 1990.
6. Rhode, 1991, 1775.
7. Dowd, 1989, 88, 91, 101 n. 76.
8. Rhode, 1991, 1772, 1774.
9. Dowd, 1989, 102 n. 81.
10. Williams, 1991, 1607–1612.
11. Littleton, 1987, 1296–1297.
12. J. Williams, 2000, 71.
13. Williams, 1991, 1617.
14. Arden, 1998.
15. Maccoby, 1998, 65–66; McCreary, 1994.
16. Farrell, 1993.
17. Storey et al., 2000.
18. Townsend, 1989, 249.
19. Townsend, 1989, 247.

Chapter 11. A Thumb on the Scales

1. Browne, 1996.
2. *Johnson v. Santa Clara Transportation Agency,* 480 U.S. 616, 632 (1987).
3. Browne, 1997b.
4. Kelley and Michela, 1980.
5. Heilman, Block, and Lucas, 1992.

6. Heilman, Block, and Stathatos, 1997.
7. Heilman, Simon, and Repper, 1987; Heilman, 1996.
8. Heilman, Lucas, and Kaplow, 1990.
9. Heilman and Alcott, 2001.
10. Heilman, Rivero, and Brett, 1991.
11. Heilman et al., 1998.
12. Heilman, Rivero, and Brett, 1991.
13. Kellogg, 2001.
14. Frehill, 1997.
15. McSherry, 2000.
16. Dresselhaus, Franz, and Clark, 1994, 1393.
17. Lubinski, 2000, 429.
18. Dresselhaus, Franz, and Clark, 1994.
19. Wulf, 1999.
20. Holloway, 1993, 96, 100.
21. Benbow and Lubinski, 1993.
22. Wulf, 1999.
23. Hrdy, 1999, 53.
24. National Science Foundation, 1999, 85.
25. Flynn, 2000.
26. 42 U.S.C. §2000e-2(k)(1).
27. 42 U.S.C. §2000e-2(a)(1).
28. *Dothard v. Rawlinson,* 433 U.S. 321 (1977) (women); *Griggs v. Duke Power Co.,* 401 U.S. 424 (1971) (blacks).
29. 42 U.S.C. §2000e-2(k)(1)(A)(i).
30. *Lanning v. Southeastern Pennsylvania Transportation Authority,* 84 Fair Employment Practice Cases (BNA) 1012 (2000).
31. Flynn, 2000.
32. Browne, 1993b.
33. *Teamsters v. United States,* 431 U.S. 324, 335–340 (1977).
34. *Teamsters v. United States,* 431 U.S. 324, 339–340 n. 20 (1977).
35. Browne, 1999, 1993a.
36. 828 F.2d 1260 (8th Cir. 1987).
37. *EEOC v. Sears, Roebuck and Co.,* 839 F.2d 302 (7th Cir. 1988).
38. Hadfield, 1993.
39. Schultz and Petterson, 1992.
40. Williams, 1989, 814.
41. Haskell and Levinson, 1988.
42. Schultz, 1990.
43. Browne, 1989.
44. Rhoads, 1993, 27.
45. Nelson, Opton, and Wilson, 1980.
46. Paul, 1989, 30.
47. Burr, 1986.
48. Aaron and Lougy, 1986, 28.
49. *Daily Labor Report,* 1986.
50. Blumrosen, 1986, 129 n.80.
51. Rhoads, 1993, 223.
52. Rhoads, 1993, 221.
53. Rhoads, 1993, 224.
54. Rhoads, 1993, 100–103, 112–113.
55. Browne, 1989.
56. Rhoads, 1993, 232.
57. Browne, 1993a, Browne, 1989.
58. *Smith v. Virginia Commonwealth University,* 84 F.3d 672 (4th Cir. 1996).
59. 84 F.3d at 684.
60. MIT Committee on Women Faculty in the School of Science, 1999.
61. *New York Times,* March 28, 1999.
62. Kleinfeld, 1999.
63. Hausman and Steiger, 2001.

Chapter 12. Mitigating Work/Family Conflict

1. Coverman, 1983.
2. Schor, 1991.
3. Juster and Stafford, 1991; Robinson and Bostrom, 1994.
4. Juster and Stafford, 1991.
5. Juster and Stafford, 1991; Robinson and Bostrom, 1994.
6. Robinson and Bostrom, 1994.
7. Robinson and Godbey, 1999.
8. Oldham, 2000.
9. Robinson and Godbey, 1999, 85–86.
10. Juster and Stafford, 1991.
11. Press and Townsley, 1998.
12. Juster and Stafford, 1991.
13. Barnett and Rivers, 1996, 175–178;

Ferree, 1991; Juster and Stafford, 1991.

14. Hochschild, 1989.
15. Shelton, 1992.
16. Coverman, 1983; Hersch, 1991; Hersch and Stratton, 1994.
17. Brines, 1994.
18. Hersch and Stratton, 1994, 123.
19. South and Spitze, 1994. Also Shelton and John, 1993.
20. Hall, 1992.
21. Roback, 1993, 129.
22. Levine, 1990.
23. Ferree, 1991.
24. South and Spitze, 1994.
25. Bianchi and Spain, 1996; Cowan, 1989.
26. Willingham and Cole, 1997, 147–148.
27. Rhode, 1991.
28. Okin, 1989, 181–182.
29. Ferree, 1991.
30. Chodorow, 1978, 215.
31. Scales, 1980–1981, 441.
32. Barnett and Rivers, 1996, 226–227; Young, 1999, 56–57.
33. Haas, 1992, 98.
34. Amato and Rivera, 1999; Pleck, 1997; Popenoe, 1996.
35. Popenoe, 1996, 211–212.
36. Mahony, 1995.
37. Bernard, 1981.
38. Crowley, 1998; Heckert et al., 1998; Kessler and McRae, 1982; Staines, Pottick, and Fudge, 1986.
39. Jacobsen and Rayack, 1996.
40. Stanley, Hunt, and Hunt, 1986.
41. Jacobsen and Rayack, 1996.
42. Haney, 1999.
43. Lyons, 1997.
44. Lyons, 1997, 61.
45. Hakim, 1995.
46. Barker, 1993.
47. Lettau, 1994.
48. Blank, 1990.
49. Lyons, 1997, 52–54.
50. Gruber, 1994.
51. J. Williams, 2000, 86. Also Schwartz, 1992, 194.
52. Bryson, 1993.
53. Schor, 1994, 159.
54. Carr et al., 1998.
55. Bureau of Labor Statistics, 1999.
56. Williams, 1991, 1620.
57. Joshi, Paci, and Waldfogel, 1999.
58. Connelly, 1991.
59. Anker, 1998.
60. *European Industrial Relations Review,* 1995.
61. *Nordic Business Report,* 2000.
62. Waldfogel, 1998.
63. Joshi, Paci, and Waldfogel, 1999.
64. *Economist,* 1998.
65. Hyde et al., 1995.
66. Corter and Fleming, 1995.
67. Malin, 1998.
68. Haas, 1992, 178.
69. Haas, 1992, 69.
70. Haas, 1992, 225.
71. Haas, 1992, 93, 97.
72. Haas, 1992, 140.
73. Haas, 1992, 179.
74. Selmi, 2000, 708.
75. Graglia, 1998.
76. Browne, 2001a.
77. Hattiangadi, 2000.
78. *Eurostat,* 1999.
79. *Eurostat,* 2001.
80. Hattiangadi, 2000.
81. Haas, 1992, 33–34.
82. Patterson, 2000.
83. Haas 1992; Hattiangadi, 2000.
84. Summers, 1989.
85. Ruhm, 1998.
86. Gruber, 1994.
87. Hattiangadi, 2000.
88. *Economist,* 1998; Hoem and Hoem, 1989.
89. Haas, 1992, 173–174.
90. Popenoe, 1991.
91. C. J. Williams, 2000.
92. Buttner and Lutz, 1990; Whittington et al., 1990; Zhang, Quan, and Van Meerbergen, 1994.

Part V. Sex and the Workplace

Chapter 13. Sexual Harassment

1. Browne, 1997a.
2. 42 U.S.C. § 2000e–2(a)(1).
3. Browne, 1997a; Paul, 1990.
4. *Meritor Savings Bank v. Vinson,* 477 U.S. 57, 65 (1986).
5. Terpstra and Cook, 1985.
6. Bowman, 1999; Gutek, 1985, 158.
7. Fitzgerald, 1993, 1071.
8. *Harris v. Forklift Systems, Inc.,* 510 U.S. 17 (1993).
9. Browne, 1997a.
10. Adler and Peirce, 1993.
11. *Radtke v. Everett,* 501 N.W.2d 155, 167 (Mich. 1993).
12. *Ellison v. Brady,* 924 F.2d 872, 878–879 (9th Cir. 1991).
13. Blumenthal, 1998; Corr and Jackson, 2001; Rotundo, Nguyen, and Sackett, 2001.
14. Gutek, 1992, 346.
15. Gutek, 1985, 96.
16. Abbey, 1982. Also Abbey, 1987; Shotland and Craig, 1988.
17. Workman and Johnson, 1991.
18. Buss, 1994, 145.
19. Abbey, 1982.
20. Semonsky and Rosenfeld, 1994, 515.
21. Muehlenhard and McCoy, 1991. Also Muehlenhard and Hollabaugh, 1988.
22. Mealey, 1992, 397.
23. Thornhill, 1996, 92–93.
24. Struckman-Johnson and Struckman-Johnson, 1994.
25. 924 F.2d 872 (9th Cir. 1991).
26. Hadfield, 1995, 1182.
27. Gutek, 1992, 346.
28. Brooks and Perot, 1991; Pryor and Day, 405; Schneider, 1982.
29. Wiener et al., 1995.
30. Struckman-Johnson and Struckman-Johnson, 1993.
31. Patai, 1998.
32. George, 1993, 23.
33. Hadfield, 1995, 1186.
34. Abbey, 1987, 193.
35. Gutek, Morasch, and Cohen, 1983; Stockdale, 1993.
36. Lim and Roloff, 1999.
37. Abbey, 1987; Shotland and Craig, 1988.
38. Oshige, 1995, 578.
39. Ehrenreich, 1990, 1208 n.114.
40. Chung and Asher, 1996.
41. Josefowitz and Gadon, 1989.
42. Browne, 2001a.
43. Browne, 1991, 1995b.
44. *Fox v. Sierra Development Co.,* 876 F. Supp. 1169, 1175 (D. Nev. 1995).
45. Kite and Whitley, 1996.
46. Struckman-Johnson and Struckman-Johnson, 1993.
47. Dressler, 1995.
48. French, 1995; Garvey, 1995; Jerome and Weinstein, 1996.
49. Struckman-Johnson and Struckman-Johnson, 1993.
50. Bravo and Cassedy, 1992. Also Goleman, 1991; Korn, 1993, 1386; Lutner, 1993, 624.
51. Fitzgerald, 1992, 1399.
52. Avner, 1994, 58.
53. Gutek, 1985, 54.
54. Workman and Johnson, 1991, 766.
55. Gutek, 1985, 55.
56. Clutton-Brock and Parker, 1995.
57. Betzig, 1986, 88.
58. Browne, 1997a.
59. 477 U.S. 57 (1986).
60. Tangri, Burt, and Johnson, 1982.
61. Hadfield, 1995, 1153 n. 2; Terpstra and Baker, 1986, 19.
62. Franke, 1995, 91.
63. Tangri, Burt, and Johnson, 1982, 36.
64. Studd and Gattiker, 1991.
65. Equal Employment Opportunity Commission, 2001.
66. Terpstra and Cook, 1985.
67. Studd and Gattiker, 1991, 274.
68. Tangri, Burt, and Johnson, 1982, 52.
69. Schwab, 1993, 20.

70. Fitzgerald and Weitzman, 1990; Tangri, Burt, and Johnson, 1982.
71. Browne, 1997, 48 n. 227, 52–53 nn. 247–248.
72. *Huebschen v. Dep't of Health and Soc. Servs.*, 716 F.2d 1167, 1172 (7th Cir. 1983).
73. *Gastineau v. Home Depot USA, Inc.*, No. CA 94–16831, 1996 U.S. App. LEXIS 1665 (9th Circuit).
74. *EEOC v. Domino's Pizza, Inc.*, 909 F. Supp. 1529 (M.D. Fla. 1995).
75. Terpstra and Baker, 1986.
76. Gutek, 1992, 351.
77. Clinton, 1982, 199–222.
78. White, 1985, 30.
79. Genovese, 1976, 416.
80. Jacobs, 1861, 46.
81. King, 1996, 307.
82. Genovese, 1976, 416.
83. Farley, 1978, xvi.
84. Lafontaine and Tredeau, 1986, 439.
85. Gutek, 1985, 57.
86. Laumann et al., 1994, 212–216.
87. Buss, 1996.
88. Gutek, 1992, 352.
89. Gutek, 1985, 17.
90. Gutek, 1992, 353.
91. Quinn, 1977, 36.
92. Gutek, 1985, 65.
93. Yount, 1991.
94. Kenrick, Trost, and Sheets, 1996, 29–30.
95. Kenrick, Trost, and Sheets, 1996, 48.
96. Roth, 1969, 128.
97. Townsend, 1999.
98. Bargh and Raymond, 1995, 87.
99. Bargh and Raymond, 1995, 92.
100. Dabbs, 2000, 10.
101. Pryor, LaVite, and Stoller, 1993.
102. *Lehmann v. Toys 'R' Us, Inc.*, 626 A.2d 445, 454 (1993).
103. Taylor, 1999, 122.
104. *EEOC v. Grinnell Corp.*, 63 Fair Empl. Prac. Cas. (BNA) 387 (D. Kan. 1993).
105. *Weinsheimer v. Rockwell Int'l Corp.*, 754 F. Supp. 1559, 1561 (M.D. Fla. 1990), aff'd, 949 F.2d 1162 (11th Cir. 1991).
106. Browne, 1997, 66–67.
107. *Arnold v. Seminole*, 614 F. Supp. 853 (E.D. Okla. 1985).
108. 25 F.3d 1459 (9th Cir. 1994).
109. Preston and Stanley, 1987.
110. Gutek, 1985, 32.
111. Low, 1992.
112. Josefowitz and Gadon, 1989, 22.
113. Josefowitz and Gadon, 1989, 25.
114. Josefowitz and Gadon, 1989.
115. Fitzgerald, 1993, 1071.
116. Quinn, 1977, 30.
117. Schneider, 1984, 444.
118. Gutek, Morasch, and Cohen, 1983; Schneider, 1984.
119. Tiger, 1997. Also Browne, 2001a.
120. Browne, 2001b, 1995b, 1991.

Chapter 14. Conclusion

1. McCauley, 1995.
2. Cejka and Eagly, 1999.

Bibliography

Aaron, H. J., and C. Lougy. 1986. *The Comparable Worth Controversy*. Washington, D.C.: Brookings Institution.

Abbey, A. 1982. Sex differences in attributions for friendly behavior: Do males misperceive females' friendliness? *Journal of Personality and Social Psychology* 42: 830–838.

Abbey, A. 1987. Misperceptions of friendly behavior as sexual interest: A survey of naturally occurring incidents. *Psychology of Women Quarterly* 11:173–194.

Adair, R. K. 1991. Using quantitative measures to predict persistence in the natural sciences. *College and University* 67:73–79.

Adams, G. R., M. Summers, and V. A. Christopherson. 1993. Age and gender differences in preschool children's identification of the emotions of others: A brief report. *Canadian Journal of Behavioural Science* 25:97–107.

Adler, R. S., and E. R. Peirce. 1993. The legal, ethical, and social implications of the "reasonable woman" standard in sexual harassment cases. *Fordham Law Review* 61:773–827.

Ahlgren, A. 1983. Sex differences in the correlates of cooperative and competitive school attitudes. *Developmental Psychology* 19:881–888.

Ahlgren, A., and D. W. Johnson. 1979. Sex differences in cooperative and competitive attitudes from the 2nd through the 12th grades. *Developmental Psychology* 15: 45–49.

Alexander, R. D. 1979. *Darwinism and Human Affairs*. Seattle: University of Washington Press.

Alington, D. E., and R. C. Leaf. 1991. Elimination of SAT-Verbal sex differences was due to policy-guided changes in item content. *Psychological Reports* 68:541–542.

Allee, W. C., N. E. Collias, and C. Z. Lutherman. 1939. Modification of the social order in flocks of hens by the injection of testosterone propionate. *Physiological Zoology* 12:412–440.

Alper, J. 1993. The pipeline is leaking women all the way along. *Science* 260:409–411.

Altermatt, E. R., J. Jovanovic, and M. Perry. 1998. Bias or responsivity? Sex and achievement-level effects on teachers' classroom questioning practices. *Journal of Educational Psychology* 90:516–527.

Amato, P. R., and A. Booth. 1996. A prospective study of divorce and parent-child relationships. *Journal of Marriage and Family* 58:356–365.

Amato, P. R., and F. Rivera. 1999. Paternal involvement and children's behavior problems. *Journal of Marriage and Family* 61:375–384.

American Association of Colleges of Pharmacy. 2001. *Pharmacy Education.* <www.aacp.org/students/pharmacyeducation.html>.

American Association of Nurse Anesthetists. 2001. *What Is a Nurse Anesthetist?* <www.aana.com/about/ataglance.asp>.

Anker, R. 1998. *Gender and Jobs: Sex Segregation of Occupations in the World.* Geneva: International Labour Office.

Arch, E. C. 1993. Risk-taking: A motivational basis for sex differences. *Psychological Reports* 73:3–11.

Archer, J. 1991. The influence of testosterone on human aggression. *British Journal of Psychology* 82:1–28.

Arden, R. 1998. Sex at work. *Prospect,* October, 16–20.

Argote, L. M., J. E. Fisher, P. J. McDonald, and E. C. O'Neal. 1976. Competitiveness in males and in females: Situational determinants of fear of success behavior. *Sex Roles* 2:295–303.

Arnold, A. P. 1996. Genetically triggered sexual differentiation of brain and behavior. *Hormones and Behavior* 30:495–505.

Aros, J. R., G. A. Henly, and N. T. Curtis. 1998. Occupational sextype and sex differences in vocational preference-measured interest relationships. *Journal of Vocational Behavior* 53:227–242.

Astin, H. S. 1991. Citation classics: Women's and men's perceptions of their contributions to science. In H. Zuckerman, J. R. Cole, and J. T. Bruer (eds.), *The Outer Circle: Women in the Scientific Community.* New York: W. W. Norton.

Atkinson, J. W. 1957. Motivational determinants of risk-taking behavior. *Psychological Review* 64:359–372.

Atkinson, J. W., J. R. Bastian, R. W. Earl, and G. H. Litwin. 1960. The achievement motive, goal setting, and probability preferences. *Journal of Abnormal and Social Psychology* 60:27–36.

Avner, J. I. 1994. Sexual harassment: Building a consensus for change. *Kansas Journal of Law and Public Policy* 3:57–76.

Baber, K. M., and P. Monaghan. 1988. College women's career and motherhood expectations: New options, old dilemmas. *Sex Roles* 19:189–203.

Backman, M. E. 1979. Patterns of mental abilities of adolescent males and females from different ethnic and socioeconomic backgrounds. In L. Willerman and R. G. Turner (eds.), *Readings about Individual and Group Differences.* San Francisco: Freeman.

Badger, K., R. S. Craft, and L. C. Jensen. 1998. Age and gender differences in value orientation among American adolescents. *Adolescence* 33:591–596.

Baenninger, M., and N. Newcombe. 1989. The role of experience in spatial test performance: A meta-analysis. *Sex Roles* 20:327–344.

Baker, J. G. 1998. Gender, race, and Ph.D. completion in natural science and engineering. *Economics of Education Review* 17:179–188.

Balkin, J. 1988. Why policemen don't like policewomen. *Journal of Police Science and Administration* 16:29–38.

Bargh, J. A., and P. Raymond. 1995. The naive misuse of power: Nonconscious sources of sexual harassment. *Journal of Social Issues* 51:85–96.

Barinaga, M. 1994. Surprises across the cultural divide. *Science* 263:1468–1472.

Barker, K. 1993. Changing assumptions and contingent solutions: The costs and benefits of women working full- and part-time. *Sex Roles* 28:47–71.

Barnett, R. C., P. Carr, A. D. Boisnier, A. Ash, R. H. Friedman, M. A. Moskowitz, and L. Szalacha. 1998. Relationships of gender and career motivation to medical faculty members' production of academic publications. *Academic Medicine* 73:180–186.

Barnett, R. C., and C. Rivers. 1996. *She Works He Works: How Two-Income Families Are Happy, Healthy, and Thriving.* Cambridge, Mass.: Harvard University Press.

Barnfield, A.M.C. 1999. Development of sex differences in spatial memory. *Perceptual and Motor Skills* 89:339–350.

Bartlett, R. L., J. H. Grant, and T. I. Miller. 1990. Personality differences and executive compensation. *Eastern Economic Journal* 16:187–195.

Bartlett, R. L., and T. I. Miller. 1985. Executive compensation: Female executives and networking. *American Economic Review* 75:266–270.

Baumeister, R. F., K. R. Catanese, and K. D. Vohs. 2001. Is there a gender difference in strength of sex drive? Theoretical views, conceptual distinctions, and a review of relevant evidence. *Personality and Social Psychology Review* 5:242–273.

Becker, G. S. 1985. Human capital, effort, and the sexual division of labor. *Journal of Labor Economics* 3:S33–S58.

Begum, H. A., and E. Ahmed. 1986. Individual risk taking and risky shift as a function of cooperation-competition proneness of subjects. *Psychological Studies* 31: 21–25.

Bell, R. Q., and L. V. Harper. 1977. *Child Effects on Adults.* Mahwah, N.J.: Erlbaum.

Bellas, M. L. 1994. Comparable worth in academia: The effects on faculty salaries of the sex composition and labor-market conditions of academic disciplines. *American Sociological Review* 59:807–821.

Benbow, C. P. 1988. Sex differences in mathematical reasoning ability in intellectually talented preadolescents: Their nature, effects, and possible causes. *Behavioral and Brain Sciences* 11:169–232.

Benbow, C. P., and D. Lubinski. 1993. Psychological profiles of the mathematically talented: Some sex differences and evidence supporting their biological basis. In *The Origins and Development of High Ability.* Wiley, Chichester (Ciba Foundation Symposium).

Benbow, C. P., and J. C. Stanley. 1996. Inequity in equity: How "equity" can lead to inequity for high-potential students. *Psychology, Public Policy, and Law* 2:249–292.

Benenson, J. F. 1999. Females' desire for status cannot be measured using male definitions. *Behavioral and Brain Sciences* 22:216–217.

Benenson, J. F., R. Del Bianco, M. Philippoussis, and N. H. Apostoleris. 1997. Girls' expression of their own perspectives in the presence of varying numbers of boys. *International Journal of Behavioral Development* 21:389–405.

Benenson, J. F., M. Philippoussis, and R. Leeb. 1999. Sex differences in neonates' cuddliness. *Journal of Genetic Psychology* 160:332–342.

Berenbaum, S. A. 1999. Effects of early androgens on sex-typed activities and interests in adolescents with congenital adrenal hyperplasia. *Hormones & Behavior* 35: 102–110.

Berenbaum, S. A., and M. Hines. 1992. Early androgens are related to childhood sex-typed toy preferences. *Psychological Science* 3:203–206.

Berenbaum, S. A., and S. M. Resnick. 1997. Early androgen effects on aggression in children and adults with congenital adrenal hyperplasia. *Psychoneuroendocrinology* 22:505–515.

Berenbaum, S. A., and E. Snyder. 1995. Early hormonal influences on childhood sex-typed activity and playmate preferences: Implications for the development of sexual orientation. *Developmental Psychology* 31:31–42.

Bergen, D. J., and J. E. Williams. 1991. Sex stereotypes in the United States revisited: 1972–1988. *Sex Roles* 24:413–423.

Berglund, A. 1993. Risky sex: Male pipefishes mate at random in the presence of a predator. *Animal Behaviour* 46:169–175.

Bernard, J. 1981. The good-provider role: Its rise and fall. *American Psychologist* 36: 1–12.

Bernard, J. 1945. Observation and generalization in cultural anthropology. *American Journal of Sociology* 50:284–291.

Bettencourt, B. A., and N. Miller. 1996. Gender differences in aggression as a function of provocation: A meta-analysis. *Psychological Bulletin* 119:422–447.

Betz, M., and L. O'Connell. 1987. Gender and work: A look at sex differences among pharmacy students. *American Journal of Pharmaceutical Education* 51:39–43.

Betzig, L. 1993. Sex, succession, and stratification in the first six civilizations: How powerful men reproduced, passed power on to their sons, and used power to defend their wealth, women, and children. In L. Ellis (ed.), *Social Stratification and Socioeconomic Inequality,* vol. 1. Westport, Conn.: Praeger.

Betzig, L. 1986. *Despotism and Differential Reproduction: A Darwinian View of History.* New York: Aldine.

Bianchi, S. M., and D. Spain. 1996. Women, work, and family in America. *Population Bulletin* 51:1–48.

Bigler, R. S., and L. S. Liben. 1990. The role of attitudes and interventions in gender-schematic processing. *Child Development* 61:1440–1452.

Bishop, P., K. Cureton, and M. Collins. 1987. Sex difference in muscular strength in equally trained men and women. *Ergonomics* 30:675–687.

Bjorklund, D. F., and K. Kipp. 1996. Parental investment theory and gender differences in the evolution of inhibition mechanisms. *Psychological Bulletin* 120:163–188.

Blakemore, J.E.O., A. A. LaRue, and A. B. Olejnik. 1979. Sex-appropriate toy preference and the ability to conceptualize toys as sex-role-related. *Developmental Psychology* 15:339–340.

Blank, R. 1990. Are part-time jobs bad jobs? In G. Burtless (ed.), *A Future of Lousy Jobs?: The Changing Structure of U.S. Wages.* Washington, D.C.: Brookings Institution.

Blau, F. D., and L. M. Kahn. 1992. The gender earnings gap: Learning from international comparisons. *American Economic Review* 82(2): 533–538.

Blumenthal, J. A. 1998. The reasonable woman standard: A meta-analytic review of gender differences in perceptions of sexual harassment. *Law and Human Behavior* 22:33–57.

Blumrosen, R. 1986. Remedies for wage discrimination. *University of Michigan Journal of Law Reform* 20:99–161.

Bonner, J. T. 1980. *The Evolution of Culture in Animals.* Princeton, N.J.: Princeton University Press.

Bowler, M. 1999. Women's earnings: An overview. *Monthly Labor Review* 122(12): 13–21.

Bowman, K. 1999. Sexual harassment in the workplace: A minor problem. *Roll Call,* March 25.

Bradley, H. 1989. *Men's Work, Women's Work: A Sociological History of the Sexual Division of Labor in Employment.* Minneapolis: University of Minnesota Press.

Bravo, E., and E. Cassedy. 1992. *The 9 to 5 Guide to Combating Sexual Harassment.* New York: Wiley.

Breedlove, S. M. 1994. Sexual differentiation of the human nervous system. *Annual Review of Psychology* 45:389–418.

Breedlove, S. M., B. M. Cooke, and C. L. Jordan. 1999. The orthodox view of brain sexual differentiation. *Brain, Behavior and Evolution* 54:8–14.

Brenner, O. C., and T. M. Bertsch. 1983. Do assertive people prefer merit pay? *Psychological Reports* 52:595–598.

Bretz, R. D., Jr., and T. A. Judge. 1994. Person-organization fit and the theory of work adjustment: Implications for satisfaction, tenure, and career success. *Journal of Vocational Behavior* 44:32–54.

Brewer, D. D., J. J. Potterat, S. B. Garrett, S. Q. Muth, J. M. Roberts Jr., D. Kasprzyk, D. E. Montano, and W. W. Darrow. 2000. Prostitution and the sex discrepancy in reported number of sexual partners. *Proceedings of the National Academy of Sciences* 97:12385–12388.

Bridges, J. S. 1989. Sex differences in occupational values. *Sex Roles* 20:205–211.

Brines, J. 1994. Economic dependency, gender, and the division of labor at home. *American Journal of Sociology* 100:652–688.

Brooks, L., and A. R. Perot. 1991. Reporting sexual harassment: Exploring a predictive model. *Psychology of Women Quarterly* 15:31–47.

Brophy, D. J. 1989. Financing the women-owned entrepreneurial firm. In O. Hagan, C. Rivchun, and D. Sexton (eds.), *Women-Owned Businesses*. New York: Praeger.

Brown, D. E. 1991. *Human Universals*. New York: McGraw-Hill.

Brown, J. P., and J. D. Silverman. 1999. The current and future market for veterinarians and veterinary medical services in the United States: Executive summary. *Journal of the American Veterinary Medical Association* 215:161–183.

Browne, K. R. 2001a. Women at war: An evolutionary perspective. *Buffalo Law Review* 49:51–247.

Browne, K. R. 2001b. Zero tolerance for the First Amendment: Title VII's regulation of employee speech. *Ohio Northern University Law Review* 27:563–605.

Browne, K. R. 1999. The use and abuse of statistical evidence in discrimination cases. In T. Loenen and P. R. Rodrigues (eds.), *Non-Discrimination Law: Comparative Perspectives*. The Hague: Kluwer.

Browne, K. R. 1998a. *Divided Labours: An Evolutionary View of Women at Work*. New Haven, Conn.: Yale University Press.

Browne, K. R. 1998b. An evolutionary account of women's workplace status. *Managerial and Decision Economics* 19:427–440.

Browne, K. R. 1997a. An evolutionary perspective on sexual harassment: Seeking roots in biology rather than ideology. *Journal of Contemporary Legal Issues* 8: 5–77.

Browne, K. R. 1997b. Nonremedial justifications for affirmative action in employment: A critique of the Justice Department position. *Labor Lawyer* 12:451–473.

Browne, K. R. 1996. Affirmative action: A rose by any other name. *Ohio Northern University Law Review* 22:1125–1157.

Browne, K. R. 1995a. Sex and temperament in modern society: A Darwinian view of the glass ceiling and the gender gap. *Arizona Law Review* 37:971–1106.

Browne, K. R. 1995b. Workplace censorship: A response to Professor Sangree. *Rutgers Law Review* 47:579–594.

Browne, K. R. 1993a. Statistical proof of discrimination: Beyond "damned lies." *Washington Law Review* 68:477–558.

Browne, K. R. 1993b. The Civil Rights Act of 1991: A "quota bill," a codification of *Griggs*, a partial return to *Wards Cove*, or all of the above? *Case Western Reserve Law Review* 43:287–400.

Browne, K. R. 1991. Title VII as censorship: Hostile environment harassment and the

First Amendment. *Ohio State Law Journal* 52:481–550.

Browne, K. R. 1989. Comparable worth: An impermissible form of affirmative action? *Loyola of Los Angeles Law Review* 22:717–759.

Browne, K. R. 1984. Biology, equality, and the law: The legal significance of biological sex differences. *Southwestern Law Journal* 38:617–702.

Bryson, L. 1993. Equality, parenting and policy making. *Australian Journal of Marriage and Family* 14:66–75.

Bureau of Labor Statistics. 1999. *Employment Characteristics of Families in 1999.* <ftp://146.142.4.23/pub/news.release/famee.txt>.

Burke, P. J. 1989. Gender identity, sex, and school performance. *Social Psychology Quarterly* 52:159–169.

Burke, R. J. 1998. Dual career couples: Are men still advantaged? *Psychological Reports* 82:209–210.

Burr, R. E. 1986. Rank injustice. *Policy Review* 38 (fall): 73–74.

Buss, D. M. 2000. *The Dangerous Passion: Why Jealousy Is as Necessary as Love and Sex.* New York: Free Press.

Buss, D. M. 1996. Sexual conflict: Evolutionary insights into feminism and the "battle of the sexes." In D. M. Buss and N. M. Malamuth (eds.), *Sex, Power, Conflict: Evolutionary and Feminist Perspectives.* New York: Oxford University Press.

Buss, D. M. 1994. *The Evolution of Desire: Strategies of Human Mating.* New York: Basic.

Buss, D. M., and D. P. Schmitt. 1993. Sexual strategies theory: An evolutionary perspective on human mating. *Psychological Review* 100:204–232.

Buss, D. M., T. K. Shackelford, L. A. Kirkpatrick, and R. J. Larsen. 2001. A half century of mate preferences: The cultural evolution of values. *Journal of Marriage and Family* 63: 491–503.

Buttner, E. H., and B. Rosen. 1988. Bank loan officers' perceptions of the characteristics of men, women, and successful entrepreneurs. *Journal of Business Venturing* 3:249–258.

Buttner, T., and W. Lutz. 1990. Estimating fertility responses to policy measures in the German Democratic Republic. *Population and Development Review* 16: 539–555.

Byrnes, J. P., D. C. Miller, and W. D. Schafer. 1999. Gender differences in risk taking: A meta-analysis. *Psychological Bulletin* 125:367–383.

Caldera, Y. M., A. C. Huston, and M. O'Brien. 1989. Social interactions and play patterns of parents and toddlers with feminine, masculine, and neutral toys. *Child Development* 60:70–76.

Callahan-Levy, C. M., and L. A. Messé. 1979. Sex differences in the allocation of pay. *Journal of Personality and Social Psychology* 37:433–446.

Campbell, A. 1999. Staying alive: Evolution, culture, and women's intrasexual aggression. *Behavioral and Brain Sciences* 22:203–252.

Campbell, A., and S. Muncer. 1994. Sex differences in aggression: Social representation and social roles. *British Journal of Social Psychology* 33:233–240.

Campbell, A., S. Muncer, and J. Odber. 1998. Primacy of organising effects of testosterone. *Behavioral and Brain Sciences* 21:365–365.

Campbell, A., L. Shirley, C. Heywood, and C. Crook. 2000. Infants' visual preference for sex-congruent babies, children, toys and activities: A longitudinal study. *British Journal of Developmental Psychology* 18:479–498.

Campenni, C. E. 1999. Gender stereotyping of children's toys: A comparison of parents and nonparents. *Sex Roles* 40:121–138.

Carr, P. L., A. S. Ash, R. H. Friedman, A. Scaramucci, R. C. Barnett, L. Szalacha,

A. Palepu, and M. A. Moskowitz. 1998. Relation of family responsibilities and gender to the productivity and career satisfaction of medical faculty. *Annals of Internal Medicine* 129:532–538.

Carretta, T. R. 1997. Group differences on US Air Force pilot selection tests. *International Journal of Selection and Assessment* 5:115–127.

Casey, M. B., R. Nuttall, E. Pezaris, and C. P. Benbow. 1995. The influence of spatial ability on gender differences in mathematics college entrance test scores across diverse samples. *Developmental Psychology* 31: 697–705.

Cashdan, E. 1996. Hormones, sex, and status in women. *Hormones and Behavior* 29:354–366.

Cashdan, E. 1998. Are men more competitive than women? *British Journal of Social Psychology* 37:213–229.

Cejka, M. A., and A. H. Eagly. 1999. Gender-stereotypic images of occupations correspond to the sex segregation of employment. *Personality and Social Psychology Bulletin* 25:413–423.

Chamallas, M. 1992. Feminist constructions of objectivity: Multiple perspectives in sexual and racial harassment litigation. *Texas Journal of Women and the Law* 1:95–142.

Charlesworth, W. R. 1996. Co-operation and competition: Contributions to an evolutionary and developmental model. *International Journal of Behavioral Development* 19:25–38.

Charlesworth, W. R., and P. J. La Freniere. 1983. Dominance, friendship, and resource utilization in preschool children's groups. *Ethology and Sociobiology* 4:175–186.

Chauvin, K. W., and R. A. Ash. 1994. Gender earnings differentials in total pay, base pay, and contingent pay. *Industrial and Labor Relations Review* 47:634–649.

Chodorow, N. 1978. *The Reproduction of Mothering: Psychoanalysis and the Sociology of Gender.* Berkeley: University of California Press.

Christensen, S. E., and S. M. Breedlove. 1998. Seductive allure of dichotomies. *Behavioral and Brain Sciences* 21:367.

Chung, T.-Y., and S. R. Asher. 1996. Children's goals and strategies in peer conflict situations. *Merrill-Palmer Quarterly* 42:125–147.

Clark, R. D. III, and E. Hatfield. 1989. Gender differences in receptivity to sexual offers. *Journal of Psychology and Human Sexuality* 2:39–55.

Clinton, C. 1982. *The Plantation Mistress: Woman's World in the Old South.* New York: Pantheon.

Clutton-Brock, T. H. 1991. *The Evolution of Parental Care.* Princeton, N.J.: Princeton University Press.

Clutton-Brock, T. H. 1985. Reproductive success in red deer. *Scientific American* 252(2): 86–92.

Clutton-Brock, T. H., and G. A. Parker. 1995. Sexual coercion in animal societies. *Animal Behaviour* 49:1345–1365.

Cobb-Clark, D. A., and Y. Dunlop. 1999. The role of gender in job promotions. *Monthly Labor Review* 122(12): 32–38.

Cohen, D. 1998. Shaping, channeling, and distributing testosterone in social systems. *Behavioral and Brain Sciences* 21:367–368.

Colangelo, N., and B. A. Kerr. 1990. Extreme academic talent: Profiles of perfect scorers. *Journal of Educational Psychology* 82:404–409.

Colapinto, J. 2000. *As Nature Made Him: The Boy Who Was Raised as a Girl.* New York: HarperCollins.

Colarelli, S. M., R. H. Alampay, and K. G. Canali. 2001. Sex composition and influence in small groups: An evolutionary perspective and review of the literature. Unpublished manuscript.

Cole, J. R., and B. Singer. 1991. A theory of limited differences: Explaining the productivity puzzle in science. In H. Zuckerman, J. R. Cole, and J. T. Bruer (eds.), *The Outer Circle: Women in the Scientific Community.* New York: W. W. Norton.

Cole, S., and R. Fiorentine. 1991. Discrimination against women in science: The confusion of outcome with process. In H. Zuckerman, J. R. Cole, and J. T. Bruer (eds.), *The Outer Circle: Women in the Scientific Community.* New York: W. W. Norton.

College Board. 2000. *2000 SAT I Test Performance by Gender.* <http://www.collegeboard.org/sat/cbsenior/cbs/cbs00/topsrs00.html>

College Board. 1997. *Common Sense about SAT Score Differences and Test Validity.* Research Notes, RN–01, June. <http://www.collegeboard.org/research/html/rn01.pdf>.

Collier, C. M., A. R. Spokane, and J. A. Bazler. 1998. Appraising science career interests in adolescent girls and boys. *Journal of Career Assessment* 6:37–48.

Collins, D. W., and D. Kimura. 1997. A large sex difference on a two-dimensional mental rotation task. *Behavioral Neuroscience* 111:845–849.

Connelly, R. 1991. The importance of child care costs to women's decision making. In D. M. Blau (ed.), *The Economics of Child Care.* New York: Russell Sage Foundation.

Corbin, C. B. 1981. Sex of subject, sex of opponent, and opponent ability as factors affecting self-confidence in a competitive situation. *Journal of Sport Psychology* 3:265–270.

Corr, P. J., and C. J. Jackson. 2001. Dimensions of perceived sexual harassment: Effects of gender, and status/liking of protagonist. *Personality and Individual Differences* 30: 525–539.

Corter, C. M., and A. S. Fleming. 1995. Psychobiology of maternal behavior in human beings. In M. H. Bornstein (ed.), *Handbook of Parenting,* vol. 2. Hillsdale, N.J.: Erlbaum.

Costa, P. T. Jr., and R. R. McCrae. 1987. On the need for longitudinal evidence and multiple measures in behavioral-genetic studies of adult personality. *Behavioral & Brain Sciences* 10:22–23.

Coverman, S. 1983. Gender, domestic labor time, and wage inequality. *American Sociological Review* 48:623–637.

Cowan, A. L. 1989. Poll finds women's gains have taken personal toll. *New York Times,* August 21, A–1.

Cronin, C. 1995. Construct validation of the Strong Interest Inventory Adventure Scale using the Sensation Seeking Scale among female college students. *Measurement and Evaluation in Counseling and Development* 28:3–8.

Cronin, H. 1991. *The Ant and the Peacock: Altruism and Sexual Selection from Darwin to Today.* Cambridge: Cambridge University Press.

Crowley, M. S. 1998. Men's self-perceived adequacy as the family breadwinner: Implications for their psychological, marital and work-family well-being. *Journal of Family and Economic Issues* 19:7–23.

Croxton, J. S., D. Chiacchia, and C. Wagner. 1987. Gender differences in attitudes toward sports and reactions to competitive situations. *Journal of Sport Behavior* 10:167–177.

D'Andrade, R. G. 1966. Sex differences and cultural institutions. In E. E. Maccoby (ed.), *The Development of Sex Differences.* Stanford, Calif.: Stanford University Press.

Dabbs, J. M. 2000. *Heroes, Rogues, and Lovers: Testosterone and Behavior.* New York: McGraw-Hill.

Dabbs, J. M., Jr. 1990. Salivary testosterone measurements: Reliability across hours, days, and weeks. *Physiology and Behavior* 48:83–86.

Dabbs, J. M., Jr., T. S. Carr, R. L. Frady, and J. K. Riad. 1995. Testosterone, crime, and misbehavior among 692 male prison inmates. *Personality and Individual Differences* 18:627–633.

Dabbs, J. M., Jr., E.-L. Chang, R. A. Strong, and R. Milun. 1998. Spatial ability, navigation strategy, and geographic knowledge among men and women. *Evolution & Human Behavior* 19:89–98.

Dabbs, J. M., Jr., C. H. Hopper, and G. J. Jurkovic. 1990. Testosterone and personality among college students and military veterans. *Personality and Individual Differences* 11:1263–1269.

Daily Labor Report. 1986. Ohio will spend $4.5 million annually to eliminate gender bias from state jobs. *Daily Labor Report (BNA),* no. 60, p. A2, March 28.

Daly, M., and M. Wilson. 1988. *Homicide.* New York: Aldine de Gruyter.

Daly, M., and M. Wilson. 1983. *Sex, Evolution and Behavior.* 2d ed. Belmont, Calif.: Wadsworth.

Darwin, C. 1871. *The Descent of Man and Selection in Relation to Sex.* London: Murray.

Darwin, C. 1859. *On the Origin of Species.* London: Murray.

Davis, J. H. 1992. Some compelling intuitions about group consensus decisions, theoretical and empirical research, and interpersonal aggregation phenomena: Selected examples, 1950–1990. *Organizational Behavior and Human Decision Processes* 52:3–38.

Dawis, R. V., and L. H. Lofquist. 1984. *A Psychological Theory of Work Adjustment: An Individual-Differences Model and Its Applications.* Minneapolis: University of Minnesota Press.

Daymont, T. N., and P. J. Andrisani. 1984. Job preferences, college major, and the gender gap in earnings. *Journal of Human Resources* 19:408–428.

Deaux, K., and J. C. Ullman. 1982. Hard-hatted women: Reflections on blue-collar employment. In H. J. Bernardin (ed.), *Women in the Work Force.* New York: Praeger.

Deitch, C. H., and S. W. Sanderson. 1987. Geographic constraints on married women's careers. *Work and Occupations* 14:616–634.

Department of Labor, Women's Bureau. 2001. *Women's Earnings as a Percent of Men's, 1979–2000.* <http://www.dol.gov/wb/public/wb_pubs/2000.htm>.

Department of Labor, Women's Bureau. 1998. *Median Annual Earnings for Year-Round, Full-Time Workers, by Sex, in Current and Real Dollars, 1951–98.* <http://www.dol.gov/wb/public/wb_pubs/mwechart.htm>.

Devine, T. J. 1994. Characteristics of self-employed women in the United States. *Monthly Labor Review* 117(3): 20–34.

Diamond, M., and H. K. Sigmundson. 1997. Sex reassignment at birth: Long-term review and clinical implications. *Archives of Pediatric and Adolescent Medicine* 151:298–304.

Diamond, M., and W. C. Young. 1963. Differential responsiveness of pregnant and nonpregnant guinea pigs to the masculinizing action of testosterone propionate. *Endocrinology* 72:429.

DiPietro, J. A. 1981. Rough and tumble play: A function of gender. *Developmental Psychology* 17:50–58.

Dittmann, R. W., M. H. Kappes, M. E. Kappes, D. Börger, H. F. Meyer-Bahlburg,

H. Stegner, R. H. Willig, and H. Wallis. 1990b. Congenital adrenal hyperplasia: II. Gender-related behavior and attitudes in female salt-wasting and simple-virilizing patients. *Psychoneuroendocrinology* 15:421–434.

Dittmann, R. W., M. H. Kappes, M. E. Kappes, D. Börger, H. Stegner, R. H. Willig, and H. Wallis. 1990a. Congenital adrenal hyperplasia: I. Gender-related behavior and attitudes in female patients and sisters. *Psychoneuroendocrinology* 15:401–420.

Donnay, D.A.C., and F. H. Borgen. 1996. Validity, structure, and content of the 1994 Strong Interest Inventory. *Journal of Counseling Psychology* 43:275–291.

Douce, L. A., and J. C. Hansen. 1990. Willingness to take risks and college women's career choice. *Journal of Vocational Behavior* 36:258–273.

Dowd, N. E. 1989. Work and family: The gender paradox and the limitations of discrimination analysis in restructuring the workplace. *Harvard Civil Rights–Civil Liberties Law Review* 24:79–172.

Dreher, G. F., and R. A. Ash. 1990. A comparative study of mentoring among men and women in managerial, professional, and technical positions. *Journal of Applied Psychology* 75:539–546.

Dresselhaus, M. S., J. R. Franz, and B. C. Clark. 1994. United States: Interventions to increase the participation of women in physics. *Science* 263:1392–1393.

Dressler, J. 1995. When "heterosexual" men kill "homosexual" men: Reflections on provocation law, sexual advances, and the "reasonable man" standard. *Journal of Criminal Law and Criminology* 85:726–763.

Duberman, L. 1975. *The Reconstituted Family.* Chicago: Nelson-Hall.

Duncan, G. J., and B. Holmlund. 1983. Was Adam Smith right after all? Another test of the theory of compensating wage differentials. *Journal of Labor Economics* 1:366–379.

Dweck, C. S., and E. S. Bush. 1976. Sex differences in learned helplessness: I. Differential debilitation with peer and adult evaluators. *Developmental Psychology* 12:147–156.

Eagly, A. H. 1995. The science and politics of comparing women and men. *American Psychologist* 50:145–158.

Eagly, A. H., and B. T. Johnson. 1990. Gender and leadership style: A meta-analysis. *Psychological Bulletin* 108:233–256.

Eagly, A. H., S. J. Karau, J. B. Miner, and B. T. Johnson. 1994. Gender and motivation to manage in hierarchic organizations: A meta-analysis. *Leadership Quarterly* 5:135–159.

Eagly, A. H., and V. J. Steffen. 1986. Gender and aggressive behavior: A meta-analytic review of the social psychological literature. *Psychological Bulletin* 100:309–330.

Eagly, A. H., and V. J. Steffen. 1984. Gender stereotypes stem from the distribution of women and men into social roles. *Journal of Personality and Social Psychology* 46:735–754.

Eagly, A. H., and W. Wood. 1999a. The origins of sex differences in human behavior: Evolved dispositions versus social roles. *American Psychologist* 54:408–423.

Eagly, A. H., and W. Wood. 1999b. The origins of aggression sex differences: Evolved dispositions versus social roles. *Behavioral and Brain Sciences* 22:223–224.

Eaton, W. O., and L. R. Enns. 1986. Sex differences in human motor activity level. *Psychological Bulletin* 100:19–28.

Eccles, J. S. 1987. Gender roles and achievement patterns: An expectancy value perspective. In J. M. Reinisch, L. A. Rosenblum, and S. A. Sanders (eds.), *Masculinity/Femininity: Basic Perspectives.* New York: Oxford University Press.

Economist. 1998. Women and work: For better, for worse, July 18.

Ehrenreich, N. S. 1990. Pluralist myths and powerless men: The ideology of reasonableness in sexual harassment law. *Yale Law Journal* 99:1177–1234.

Eide, E. 1994. College major choice and changes in the gender wage gap. *Contemporary Economic Policy* 12:55–64.

Eisenberg, N., E. Murray, and T. Hite. 1982. Children's reasoning regarding sex-typed toy choices. *Child Development* 53:81–86.

Ellis, L. 1986. Evidence of neuroandrogenic etiology of sex roles from a combined analysis of human, nonhuman primate and nonprimate mammalian studies. *Personality and Individual Differences* 7:519–552.

England, P. 1979. Women and occupational prestige: A case of vacuous sex equality. *Signs* 5:252–265.

Equal Employment Opportunity Commission. 2001. *Sexual Harassment Charges, EEOC & FEPAs Combined: FY 1992–FY 2000.* <www.eeoc.gov/stats/harass/html>.

European Industrial Relations Review. 1995. *Parental Leave in Europe* 262:14–23.

Eurostat. 2001. Employment rates vary between 53.4% and 76.4% in the EU. *Eurostat News Release,* no. 69/2001.

Eurostat. 1999. Average EU woman earns a quarter less than a man. *Eurostat News Release,* no. 48/99.

Fagot, B. I. 1985. Beyond the reinforcement principle: Another step toward understanding sex role development. *Developmental Psychology* 21:1097–1104.

Fagot, B. I. 1984. Teacher and peer reactions to boys' and girls' play styles. *Sex Roles* 11:691–702.

Farley, L. 1978. *Sexual Shakedown: The Sexual Harassment of Women on the Job.* New York: McGraw-Hill.

Farmer, H. S. 1988. Predicting who our future scientists and mathematicians will be. *Behavioral and Brain Sciences* 11:190–191.

Farrell, W. 1993. *The Myth of Male Power: Why Men Are the Disposable Sex.* New York: Simon & Schuster.

Fausto-Sterling, A. 1993. The five sexes: Why male and female are not enough. *The Sciences* 33(2): 20–24.

Fausto-Sterling, A. 1992. *Myths of Gender: Biological Theories About Women and Men.* (2d ed.). New York: Basic.

Fein, G., D. Johnson, N. Kosson, L. Stork, and L. Wasserman. 1975. Sex stereotypes and preferences in the toy choices of 20-month-old boys and girls. *Developmental Psychology* 11:527–528.

Feingold, A. 1995. The additive effects of differences in central tendency and variability are important in comparisons between groups. *American Psychologist* 50: 5–13.

Feingold, A. 1994. Gender differences in personality: A meta-analysis. *Psychological Bulletin* 116:429–456.

Feingold, A. 1988. Cognitive gender differences are disappearing. *American Psychologist* 43:95–103.

Felmlee, D. H. 1994. Who's on top? Power in romantic relationships. *Sex Roles* 31:275–295.

Ferree, M. M. 1991. The gender division of labor in two-earner marriages: Dimensions of variability and change. *Journal of Family Issues* 12:158–180.

Fierman, J. 1990. Do women manage differently? *Fortune,* December 17, 115.

Filer, R. K. 1985. Male-female wage differences: The importance of compensating differentials. *Industrial and Labor Relations Review* 38:426–437.

Fischer, G. W. 1976. Multidimensional utility models for risky and riskless choice.

Organizational Behavior and Human Decision Processes 17:127–146.

Fisher, H. 1999. *The First Sex: The Natural Talents of Women and How They Are Changing the World.* New York: Random House.

Fisher-Thompson, D. 1993. Adult toy purchases for children: Factors affecting sex-typed toy selection. *Journal of Applied Developmental Psychology* 14:385–406.

Fitch, R. H., and V. H. Denenberg. 1998. A role for ovarian hormones in sexual differentiation of the brain. *Behavioral and Brain Sciences* 21:311–352.

Fitzgerald, L. F. 1993. Sexual harassment: Violence against women in the workplace. *American Psychologist* 48:1070–1076.

Fitzgerald, L. F. 1992. Science v. myth: The failure of reason in the Clarence Thomas hearings. *Southern California Law Review* 65:1399–1410.

Fitzgerald, L. F., and L. M. Weitzman. 1990. Men who harass: Speculation and data. In M. A. Paludi (ed.), *Ivory Power: Sexual Harassment on Campus.* Albany, N.Y.: SUNY Press.

Flinn, M. V., R. J. Quinlan, S. A. Decker, M. T. Turner, and B. G. England. 1996. Male-female differences in effects of parental absence on glucocorticoid stress response. *Human Nature* 7:125–162.

Flynn, K. 2000. Despite recruiting, few women do well in firefighter tests. *New York Times,* February 3, B–1.

Forbes, B. C. 1990. Thoughts on the business life. *Forbes,* March 5, 180.

Foss, J. 1998. Testosterone and the second sex. *Behavioral and Brain Sciences* 21:374–375.

Fox, R. 1992. Prejudice and the unfinished mind: A new look at an old failing. *Psychological Inquiry* 3:137–152.

Fox-Genovese, E. 1996. *Feminism Is Not the Story of My Life: How Today's Feminist Elite Has Lost Touch with the Real Concerns of Women.* New York: Doubleday.

Frank, R. H. 1995. What price the moral high ground? *Southern Economic Journal* 63:1–17.

Frank, R. H. 1978. Why women earn less: The theory and estimation of differential overqualification. *American Economic Review* 68:360–373.

Franke, K. M. 1995. The central mistake of sex discrimination law: The disaggregation of sex from gender. *University of Pennsylvania Law Review* 144:1–99.

Fratianni, C. M., and J. Imperato-McGinley. 1994. The syndrome of 5-alpha-reductase deficiency. *Endocrinologist* 4:302–314.

Freedman, D. G., and M. M. DeBoer. 1979. Biological and cultural differences in early child development. *Annual Review of Anthropology* 8:579–600.

Frehill, L. M. 1997. Education and occupational segregation: The decision to major in engineering. *Sociological Quarterly* 38:225–249.

French, R. 1995. Parents struggle with son's fate. *Detroit News,* December 11.

Fuchs, V. 1986. Sex differences in economic well-being. *Science* 232:459–464.

Fuchs, V. 1988. *Women's Quest for Economic Equality.* Cambridge, Mass.: Harvard University Press.

Furchtgott-Roth, D., and C. Stolba. 1999. *Women's Figures: An Illustrated Guide to the Economic Progress of Women in America.* Washington, D.C.: AEI Press.

Garton, A. F., and C. Pratt. 1991. Leisure activities of adolescent school students: Predictors of participation and interest. *Journal of Adolescence* 14:305–321.

Garvey, M. 1995. The aftershock of shock TV: Two ordinary lives were shattered in a bizarre tangle. *Washington Post,* March 25.

Gasse, Y. 1982. Elaborations on the psychology of the entrepreneur. In C. A. Kent,

D. L. Sexton, and K. H. Vesper (eds.), *Encyclopedia of Entrepreneurship*. Englewood Cliffs, N.J.: Prentice-Hall.

Gaulin, S.J.C., and R. W. FitzGerald. 1986. Sex differences in spatial ability: An evolutionary hypothesis and test. *American Naturalist* 127:74–88.

Gaulin, S.J.C., R. W. FitzGerald, and M. S. Wartell. 1990. Sex differences in spatial ability and activity in two vole species. *Journal of Comparative Psychology* 104:88–93.

Geary, D. C. 1998. *Male, Female: The Evolution of Human Sex Differences*. Washington, D.C.: American Psychological Association.

Geary, D. C. 1996. Sexual selection and sex differences in mathematical abilities. *Behavioral and Brain Sciences* 19:229–284.

Genovese, E. D. 1976. *Roll, Jordan, Roll: The World the Slaves Made*. New York: Vintage.

George, B. G. 1993. The back door: Legitimizing sexual harassment claims. *Boston University Law Review* 73:1–38.

Gerhart, B. 1990. Gender differences in current and starting salaries: The role of performance, college major, and job title. *Industrial and Labor Relations Review* 43:418–433.

Ghiselin, M. T. 1996. Differences in male and female cognitive abilities: Sexual selection or division of labor? *Behavioral and Brain Sciences* 19:254–255.

Gibbons, J. L., M. Lynn, and D. A. Stiles. 1997. Cross-national gender differences in adolescents' preferences for free-time activities. *Cross-Cultural Research* 31: 55–69.

Gibbs, A. C., and J. F. Wilson. 1999. Sex differences in route learning by children. *Perceptual and Motor Skills* 88:590–594.

Gilger, J. W. 1996. Sex differences in mathematical ability: Genes, environment, and evolution. *Behavioral and Brain Sciences* 19:255–256.

Gilligan, C. 1982. *In a Different Voice: Psychological Theory and Women's Development*. Cambridge, Mass.: Harvard University Press.

Ginsburg, H. J., and S. M. Miller. 1982. Sex differences in children's risk-taking behavior. *Child Development* 53:426–428.

Glass Ceiling Commission. 1995a. *Good for Business: Making Full Use of the Nation's Human Capital: Fact Finding Report of the Federal Glass Ceiling Commission*. Washington, D.C.: Government Printing Office.

Glass Ceiling Commission, 1995b. *A Solid Investment: Making Full Use of the Nation's Human Capital*. Washington, D.C.: Government Printing Office.

Goldberg, P. 1968. Are women prejudiced against women? *Trans-Action* 5(5): 28–32.

Goldberg, S. 1993. *Why Men Rule: A Theory of Male Dominance*. Chicago: Open Court.

Goldberg, S., S. L. Blumberg, and A. Kriger. 1982. Menarche and interest in infants: Biological and social influences. *Child Development* 53:1544–1550.

Goldberg, S., and M. Lewis. 1969. Play behavior in the year-old infant: Early sex differences. *Child Development* 40:21–31.

Goldin, C. 1990. *Understanding the Gender Gap: An Economic History of American Women*. New York: Oxford University Press.

Goldsmith, T. H. 1991. *The Biological Roots of Human Nature: Forging Links between Evolution and Behavior*. New York: Oxford University Press.

Goleman, D. 1991. Sexual harassment: It's about power not lust. *New York Times*, October 22.

Goodenough, E. W. 1957. Interest in persons as an aspect of sex difference in the early years. *Genetic Psychology Monographs* 55:287–323.

Gottfredson, G. D., and J. L. Holland. 1996. *Dictionary of Holland Occupational Codes*. 3rd ed. Odessa, Fla.: Psychological Assessment Resources.

Gottfredson, L. S. 1981. Circumscription and compromise: A developmental theory of occupational aspirations. *Journal of Counseling Psychology* 28:545–579.

Gouchie, C., and D. Kimura. 1991. The relationship between testosterone levels and cognitive ability patterns. *Psychoneuroendocrinology* 16:323–334.

Goy, R. W., F. B. Bercovitch, and M. C. McBrair. 1988. Behavioral masculinization is independent of genital masculinization in prenatally androgenized female rhesus macaques. *Hormones and Behavior* 22:552–571.

Graglia, F. C. 1998. *Domestic Tranquility: A Brief against Feminism.* Dallas, Tex.: Spence Publishing.

Grant, V. J. 1998. *Maternal Personality, Evolution and the Sex Ratio: Do Mothers Control the Sex of the Infant?* London: Routledge.

Gregersen, E. 1982. *Sexual Practices: The Story of Human Sexuality.* New York: F. Watts.

Gregersen, N. P., and H. Y. Berg. 1994. Lifestyle and accidents among young drivers. *Accident Analysis and Prevention* 26:297–303.

Grey, R. J., and G. G. Gordon. 1978. Risk-taking managers: Who gets the top jobs? *Management Review* 67(11): 8–13.

Grimshaw, G. M., G. Sitarenios, and J. K. Finegan. 1995. Mental rotation at 7 years: Relations with prenatal testosterone levels and spatial play experiences. *Brain and Cognition* 29:85–100.

Groshen, E. L. 1991. The structure of the female/male wage differential: Is it who you are, what you do, or where you work? *Journal of Human Resources* 26:457–472.

Gruber, J. 1994. The incidence of mandated maternity benefits. *American Economic Review* 84:622–641.

Gutek, B. A. 1992. Understanding sexual harassment at work. *Notre Dame Journal of Law, Ethics, and Public Policy* 6:335–358.

Gutek, B. A., B. Morasch, and A. G. Cohen. 1983. Interpreting social-sexual behavior in a work setting. *Journal of Vocational Behavior* 22:30–48.

Gutek, B. A. 1985. *Sex and the Workplace: The Impact of Sexual Behavior and Harassment on Women, Men, and Organizations.* San Francisco: Jossey-Bass.

Guttentag, M., and H. Bray. 1976. *Undoing Sex Stereotypes: Research and Resources for Educators.* New York: McGraw Hill.

Haas, L. 1992. *Equal Parenthood and Social Policy: A Study of Parental Leave in Sweden.* Albany, N.Y.: SUNY Press.

Hadfield, G. K. 1995. Rational women: A test for sex-based harassment. *California Law Review* 83:1151–1189.

Hadfield, G. K. 1993. Households at work: Beyond labor market policies to remedy the gender gap. *Georgetown Law Journal* 82:89–107.

Hakim, C. 1995. Five feminist myths about women's employment. *British Journal of Sociology* 46:429–455.

Hall, W. A. 1992. Comparison of the experience of women and men in dual-earner families following the birth of their first infant. *Image: Journal of Nursing Scholarship* 24:33–38.

Halpern, D. F. 2000. *Sex Differences in Cognitive Abilities.* 3rd ed. Mahwah, N.J.: Erlbaum.

Halpern, D. F. 1997. Sex differences in intelligence: Implications for education. *American Psychologist* 52:1091–1102.

Halpern, D. F. 1996. Mating, math achievement, and other multiple relationships. *Behavioral and Brain Sciences* 19:256.

Halpern, D. F., M. G. Haviland, and C. D. Killian. 1998. Handedness and sex differences in intelligence: Evidence from the Medical College Admission Test. *Brain and Cognition* 38:87–101.

Hamermesh, D. S. 1993. Treading water: The annual report on the economic status of the profession. *Academe* 79(2): 8–14.

Hammer, C., and R. V. Dusek. 1996. Brain differences, anthropological stories, and educational implications. *Behavioral and Brain Sciences* 19:257.

Hampson, E. 1990. Variations in sex-related cognitive abilities across the menstrual cycle. *Brain and Cognition* 14:26–43.

Hampson, E., J. F. Rovet, and D. Altmann. 1998. Spatial reasoning in children with congenital adrenal hyperplasia due to 21-hydroxylase deficiency. *Developmental Neuropsychology* 14:299–320.

Haney, D. Q. 1999. Researchers find couples working more than ever, and liking it less. *Los Angeles Times*, January 24, C–6.

Hansen, J. C. 2000. Interpretation of the Strong Interest Inventory. In Watkins, C. E., Jr., and V. L. Campbell (eds.), *Testing and Assessment in Counseling Practice*. Mahwah, N.J.: Erlbaum.

Hansen, J. C. 1988. Changing interests of women: Myth or reality? *Applied Psychology: An International Review* 37:133–150.

Haqq, C. M., C.-Y. King, E. Ukiyama, S. Falsafi, T. N. Haqq, P. K. Donahoe, and M. A. Weiss. 1994. Molecular basis of mammalian sexual determination: Activation of Müllerian Inhibiting Substance gene expression by Sry. *Science* 266:1494–1500.

Harmon, L. W., and F. H. Borgen. 1995. Advances in career assessment and the 1994 Strong Interest Inventory. *Journal of Career Assessment* 3:347–372.

Harper, L. V., and K. M. Sanders. 1975. Preschool children's use of space: Sex differences in outdoor play. *Developmental Psychology* 11:119.

Harrell, T. W. 1993. The association of marriage and MBA earnings. *Psychological Reports* 72:955–964.

Harrell, T. W., and B. Alpert. 1989. Attributes of successful MBAs: A 20-year longitudinal study. *Human Performance* 2:301–322.

Harris, J. R. 1998. *The Nurture Assumption: Why Children Turn Out the Way They Do*. New York: Free Press.

Haskell, T., and S. Levinson. 1988. Academic freedom and expert witnessing: Historians and the Sears case. *Texas Law Review* 66:1629–1659.

Hattiangadi, A. U. 2000. *Paid Family Leave: At What Cost?* Washington, D.C.: Employment Policy Foundation.

Hattiangadi, A. U., and A. M. Habib. 2000. *A Closer Look at Comparable Worth*. 2nd ed. Washington, D.C.: Employment Policy Foundation.

Hausman, P. 1999. *On the Rarity of Mathematically and Mechanically Gifted Females: A Life History Analysis*. (Ph.D. dissertation, The Fielding Institute).

Hausman, P., and J. H. Steiger. 2001. *Confession without Guilt? M.I.T. Jumped the Gun to Avoid a Sex-Discrimination Controversy, But Shot Itself in the Foot*. <www.iwf.org /news/mitfinal.pdf>.

Hawkes, K. 1991. Showing off: Tests of an hypothesis about men's foraging goals. *Ethology and Sociobiology* 12:29–54.

Hecker, D. E. 1998. Earnings of college graduates: Women compared with men. *Monthly Labor Review* 121(3): 62–71.

Heckert, D. A., T. C. Nowak, and K. A. Snyder. 1998. The impact of husbands' and wives' relative earnings on marital disruption. *Journal of Marriage and Family* 60:690–703.

Hedges, L. V., and A. Nowell. 1995. Sex differences in mental test scores, variability, and numbers of high-scoring individuals. *Science* 269:41–45.

Hedrick, A. V., and L. M. Dill. 1993. Mate choice by female crickets is influenced by predation risk. *Animal Behaviour* 46:193–196.

Heilman, M. E. 1996. Affirmative action's contradictory consequences. *Journal of Social Issues* 52:105–109.

Heilman, M. E., and V. B. Alcott. 2001. What I think you think of me: Women's reactions to being viewed as beneficiaries of preferential selection. *Journal of Applied Psychology* 86:574–582.

Heilman, M. E., W. S. Battle, C. E. Keller, and R. A. Lee. 1998. Type of affirmative action policy: A determinant of reactions to sex-based preferential selection? *Journal of Applied Psychology* 83:190–205.

Heilman, M. E., C. J. Block, and J. A. Lucas. 1992. Presumed incompetent? Stigmatization and affirmative action efforts. *Journal of Applied Psychology* 77:536–544.

Heilman, M. E., C. J. Block, and P. Stathatos. 1997. The affirmative action stigma of incompetence: Effects of performance information ambiguity. *Academy of Management Journal* 40:603–625.

Heilman, M. E., S. R. Kaplow, M. A. Amato, and P. Stathatos. 1993. When similarity is a liability: Effects of sex-based preferential selection on reactions to like-sex and different-sex others. *Journal of Applied Psychology* 78:917–927.

Heilman, M. E., J. A. Lucas, and S. R. Kaplow. 1990. Self-derogating consequences of sex-based preferential selection: The moderating role of initial self-confidence. *Organizational Behavior and Human Decision Processes* 46:202–216.

Heilman, M. E., J. C. Rivero, and J. F. Brett. 1991. Skirting the competence issue: Effects of sex-based preferential selection on task choices of women and men. *Journal of Applied Psychology* 76:99–105.

Heilman, M. E., M. C. Simon, and D. P. Repper. 1987. Intentionally favored, unintentionally harmed? The impact of sex-based preferential selection on self-perceptions and self-evaluations. *Journal of Applied Psychology* 72:62–68.

Hellerstein, J. K., and D. Neumark. 1998. Wage discrimination, segregation, and sex differences in wages and productivity within and between plants. *Industrial Relations* 37:232–260.

Hellerstein, J. K., D. Neumark, and K. R. Troske. 1996. *Wages, Productivity, and Worker Characteristics: Evidence from Plant-Level Production Functions and Wage Equations.* National Bureau of Economic Research, Working Paper 5626, June.

Hennig, M., and A. Jardim. 1977. *The Managerial Woman.* New York: Doubleday.

Hersch, J. 1995. Optimal "mismatch" and promotions. *Economic Inquiry* 33:611–624.

Hersch, J. 1991. Male-female differences in hourly wages: The role of human capital, working conditions, and housework. *Industrial and Labor Relations Review* 44:746–759.

Hersch, J., and L. S. Stratton. 1994. Housework, wages, and the division of housework time for employed spouses. *American Economic Review* 84:120–125.

Hersch, J., and W. K. Viscusi. 1996. Gender differences in promotions and wages. *Industrial Relations* 35:461–472.

Hewlett, B. S. 1988. Sexual selection and paternal investment among Aka pygmies. In L. Betzig, M. Borgerhoff Mulder, and P. Turke (eds.), *Human Reproductive Behaviour: A Darwinian Perspective.* Cambridge: Cambridge University Press.

Hier, D. B., and W. F. Crowley Jr. 1982. Spatial ability in androgen-deficient men. *New England Journal of Medicine* 306:1202–1205.

Hillier, L. M., and B. A. Morrongiello. 1998. Age and gender differences in school-age children's appraisals of injury risk. *Journal of Pediatric Psychology* 23:229–238.

Hines, M., and F. R. Kaufman. 1994. Androgen and the development of human sex-typical behavior: Rough-and-tumble play and sex of preferred playmates in children with congenital adrenal hyperplasia (CAH). *Child Development* 65:1042–1053.

Hinze, S. W. 2000. Inside medical marriages: The effect of gender on income. *Work and Occupations* 27:465–499.

Hochschild, A. 1989. *The Second Shift: Inside the Two-Job Marriage*. New York: Viking.

Hoem, B., and J. M. Hoem. 1989. The impact of women's employment on second and third births in modern Sweden. *Population Studies* 43:47–67.

Hoffman, M. L. 1977. Sex differences in empathy and related behaviors. *Psychological Bulletin* 84:712–722.

Hoffmann, C., and J. Reed. 1982. When is imbalance not discrimination? In W. Block and M. Walker (eds.), *Discrimination, Affirmative Action, and Equal Opportunity: An Economic and Social Perspective*. Vancouver: Fraser Institute.

Hoffmann, C. C., and K. P. Hoffmann. 1987. Does comparable worth obscure the real issues? *Personnel Journal* 66:83–95.

Holland, J. L. 1997. *Making Vocational Choices: A Theory of Vocational Personalities and Work Environments*. 3rd ed. Odessa, Fla.: Psychological Assessment Resources.

Holloway, M. 1993. A lab of her own. *Scientific American* 269 (November): 94–103.

Horner, M. S. 1972. Toward an understanding of achievement-related conflicts in women. *Journal of Social Issues* 28:157–175.

Horvath, P., and M. Zuckerman. 1993. Sensation seeking, risk appraisal, and risky behavior. *Personality and Individual Differences* 14:41–52.

Hoyenga, K. B., and K. T. Hoyenga. 1993. *Gender-Related Differences: Origins and Outcomes*. Boston: Allyn & Bacon.

Hrdy, S. B. 1999. *Mother Nature: A History of Mothers, Infants, and Natural Selection*. New York: Pantheon.

Hrdy, S. B. 1981. *The Woman That Never Evolved*. Cambridge, Mass.: Harvard University Press.

Huang, W. R. 1997. Gender differences in the earnings of lawyers. *Columbia Journal of Law and Social Problems* 30:267–325.

Hughes, B. J., H. J. Sullivan, and J. Beaird. 1986. Continuing motivation of boys and girls under differing evaluation conditions and achievement levels. *American Educational Research Journal* 23:660–667.

Hume, D. 1999 [1777]. *An Enquiry Concerning Human Understanding*. New York: Oxford University Press.

Humphreys, L. G., and D. Lubinski. 1996. Assessing spatial visualization: An underappreciated ability for many school and work settings. In C. P. Benbow and D. Lubinski (eds.), *Intellectual Talent: Psychometric and Social Issues*. Baltimore, Md.: Johns Hopkins University Press.

Humphreys, L. G., D. Lubinski, and G. Yao. 1993. Utility of predicting group membership and the role of spatial visualization in becoming an engineer, physical scientist, or artist. *Journal of Applied Psychology* 78:250–261.

Hundley, G. 2001. Domestic division of labor and self/organizationally employed differences in job attitudes and earnings. *Journal of Family and Economic Issues* 22:121–139.

Hunter, R. 1993. Afterword: A feminist response to the gender gap in compensation symposium. *Georgetown Law Journal* 82:147–158.

Huttenlocher, J., W. Haight, A. Bryk, M. Seltzer, and T. Lyons. 1991. Early vocabulary growth: Relation to language input and gender. *Developmental Psychology* 27:236–248.

Hyde, J. S., E. Fennema, and S. J. Lamon. 1990. Gender differences in mathematics performance: A meta-analysis. *Psychological Bulletin* 107:139–155.

Hyde, J. S., M. H. Klein, M. J. Essex, and R. Clark. 1995. Maternity leave and women's mental health. *Psychology of Women Quarterly* 19:257–285.

Iijima, M., O. Arisaka, F. Minamoto, and Y. Arai. 2001. Sex differences in children's free drawings: A study on girls with congenital adrenal hyperplasia. *Hormones and Behavior* 40:99–104.

Interagency Committee for Women's Business Enterprise. 1996. *Expanding Business Opportunities for Women.* <http://199.171.55.3/womeninbusiness/newsltr.html>.

Jacklin, C., and L. A. Baker. 1993. Early gender development. In S. Oskamp and M. Costanzo (eds.), *Gender Issues in Contemporary Society.* Claremont Symposium on Applied Social Psychology 6:41–57. Newbury Park, Calif.: Sage.

Jacklin, C. N., and E. E. Maccoby. 1978. Social behavior at thirty-three months in same-sex and mixed-sex dyads. *Child Development* 49:557–569.

Jacobs, H. 1861. *Incidents in the Life of a Slave Girl.* New York: Oxford University Press.

Jacobsen, J. P., and W. L. Rayack. 1996. Do men whose wives work really earn less? *American Economic Review* 86:268–273.

Janowsky, J. S., S. K. Oviatt, and E. S. Orwoll. 1994. Testosterone influences spatial cognition in older men. *Behavioral Neuroscience* 108:325–332.

Jensen, A. R. 1998. *The g Factor: The Science of Mental Ability.* Westport, Conn.: Praeger.

Jensen, A. R. 1981. *Straight Talk about Mental Tests.* New York: Free Press.

Jensen-Campbell, L. A., W. G. Graziano, and S. G. West. 1995. Dominance, prosocial orientation, and female preferences: Do nice guys really finish last? *Journal of Personality and Social Psychology* 68:427–440.

Jerome, R., and F. Weinstein. 1996. Playing with fire: His parents say Jenny Jones's reckless "ambush" turned their son into a killer. *People,* October 21, 58.

Johnson, E. S., and A. C. Meade. 1987. Developmental patterns of spatial ability: An early sex difference. *Child Development* 58:725–740.

Johnson, R. C. 1996. Attributes of Carnegie medalists performing acts of heroism and of the recipients of these acts. *Ethology and Sociobiology* 17:355–362.

Jones, O. D. 1999. Sex, culture, and the biology of rape: Toward explanation and prevention. *California Law Review* 87:827–941.

Jones, O. D. 1997. Evolutionary analysis in law: An introduction and application to child abuse. *North Carolina Law Review* 75:1117–1242.

Josefowitz, N., and H. Gadon. 1989. Hazing: Uncovering one of the best-kept secrets of the workplace. *Business Horizons* 32(May): 22–26.

Josephs, R. A., H. R. Markus, and R. W. Tafarodi. 1992. Gender and self-esteem. *Journal of Personality and Social Psychology* 63:391–402.

Joshi, H., P. Paci, and J. Waldfogel. 1999. The wages of motherhood: Better or worse? *Cambridge Journal of Economics* 23:543–564.

Joslyn, W. D. 1973. Androgen-induced social dominance in infant female rhesus monkeys. *Journal of Child Psychology and Psychiatry* 14:137–145.

Jussim, L. J., C. R. McCauley, and Y.-T. Lee. 1995. Why study stereotype accuracy and inaccuracy? In Y.-T. Lee, L. J. Jussim, and C. R. McCauley (eds.), *Stereotype Accuracy: Toward Appreciating Group Differences.* Washington, D.C.:American Psychological Association.

Juster, F. T., and F. P. Stafford. 1991. The allocation of time: Empirical findings, behavioral models, and problems of measurement. *Journal of Economic Literature* 29:471–522.

Kachigan, S. K. 1991. *Multivariate Statistical Analysis: A Conceptual Introduction.* 2nd ed. New York: Radius Press.

Kahn, P. 1994. Turkey: A prominent role on a stage set by history. *Science* 263:1447–1448.

Kalb, L., and L. Hugick. 1990. The American worker: How we feel about our jobs. *Public Perspective* 1(6): 21–22.

Kaplan, S. H., L. M. Sullivan, K. A. Dukes, C. F. Phillips, R. P. Kelch, and J. G. Schaller. 1996. Sex differences in academic advancement: Results of a national study of pediatricians. *New England Journal of Medicine* 335:1282–1289.

Katz, M. M., and M. J. Konner. 1981. The role of the father: An anthropological perspective. In M. E. Lamb (ed.), *The Role of the Father in Child Development.* 2nd ed. New York: Wiley.

Kaufman, A. S, and J. E. McLean. 1998. An investigation into the relationship between interests and intelligence. *Journal of Clinical Psychology* 54:279–295.

Kay, H. H. 1990. Perspectives on sociobiology, feminism, and the law. In D. L. Rhode (ed.), *Theoretical Perspectives on Sexual Difference.* New Haven, Conn.: Yale University Press.

Kelley, H. H., and J. L. Michela. 1980. Attribution theory and research. *Annual Review of Psychology* 31:457–501.

Kellogg, A. P. 2001. Report urges educators to improve the representation of women in the sciences. *Chronicle of Higher Education,* July 18. <http://chronicle.com/daily/2001/07/2001071802n.htm>.

Kenrick, D. T., M. R. Trost, and V. L. Sheets. 1996. Power, harassment, and trophy mates: The feminist advantages of an evolutionary perspective. In D. M. Buss and N. M. Malamuth (eds.), *Sex, Power, Conflict: Evolutionary and Feminist Perspectives.* New York: Oxford University Press.

Kerns, K. A., and S. A. Berenbaum. 1991. Sex differences in spatial ability in children. *Behavior Genetics* 21:383–396.

Kessler, R. C., and J. A. McRae. 1982. The effect of wives' employment on the mental health of married men and women. *American Sociological Review* 47:216–227.

Kilbourne, B. S., and P. England. 1996. Occupational skill, gender, and earnings. In P. J. Dubeck and K. Borman (eds.), *Women and Work: A Handbook.* New York: Garland.

Kim, M. K., and S. Polachek. 1994. Panel estimates of male-female earnings functions. *Journal of Human Resources* 29:406–428.

Kimball, M. M. 1996. Some problematic links between hunting and geometry. *Behavioral and Brain Sciences* 19:258–259.

Kimball, M. M. 1989. A new perspective on women's math achievement. *Psychological Bulletin* 105:198–214.

Kimura, D. 1999. *Sex and Cognition.* Cambridge, Mass.: MIT Press.

King, W. 1996. Suffer with them till death: Slave women and their children in nineteenth-century America. In D. B. Gaspar and D. C. Hine, *More than Chattel: Black Women and Slavery in the Americas.* Bloomington: Indiana University Press.

King Solomon's Mines. 1985. MGM/UA Studios.

Kirkcaldy, B., and C. L. Cooper. 1992. Work attitudes and leisure preferences: Sex differences. *Personality and Individual Differences* 13:329–334.

Kirkcaldy, B., and A. Furnham. 1993. Predictors of beliefs about money. *Psychological Reports* 73:1079–1082.

Kirshnit, C. E., M. Ham, and M. H. Richards. 1989. The sporting life: Athletic activities during early adolescence. *Journal of Youth and Adolescence* 18:601–615.

Kitcher, P. 1985. *Vaulting Ambition: Sociobiology and the Quest for Human Nature*. Cambridge, Mass.: MIT Press.

Kite, M. E., and B. E. Whitley Jr. 1996. Sex differences in attitudes toward homosexual persons, behaviors, and civil rights: A meta-analysis. *Personality and Social Psychology Bulletin* 22:336–353.

Kleinfeld, J. S. 1999. *MIT Tarnishes Its Reputation with Gender Junk Science*. <www.uaf.edu/northern/mitstudy/mittarn.pdf>.

Kleinfeld, J. S. 1998. *The Myth That Schools Shortchange Girls: Social Science in the Service of Deception*. <http://www.uaf.edu/northern/schools/myth.pdf>.

Kline, T.J.B., and Y. P. Sell. 1996. Cooperativeness vs. competitiveness: Initial findings regarding effects on the performance of individual and group problem solving. *Psychological Reports* 79:355–365.

Kohlberg, L. A. 1966. A cognitive-developmental analysis of children's sex-role concepts and attitudes. In E. E. Maccoby (ed.), *The Development of Sex Differences*. Stanford, Calif.: Stanford University Press.

Kolakowski, D., and R. M. Malina. 1974. Spatial ability, throwing accuracy and man's hunting heritage. *Nature* 251:410–412.

Konner, M., and C. Worthman. 1980. Nursing frequency, gonadal function, and birth spacing among !Kung hunter-gatherers. *Science* 207:788–791.

Konrad, A. M., J. E. Ritchie Jr., P. Lieb, and E. Corrigall. 2000. Sex differences and similarities in job attribute preferences: A meta-analysis. *Psychological Bulletin* 126:593–641.

Koot, H. M., and F. C. Verhulst. 1991. Prevalence of problem behavior in Dutch children aged 2–3. *Acta Psychiatrica Scandinavica* 83:1–37.

Korenman, S., and D. Neumark. 1994. Sources of bias in women's wage equations: Results using sibling data. *Journal of Human Resources* 29:379–405.

Korenman, S., and D. Neumark. 1992. Marriage, motherhood, and wages. *Journal of Human Resources* 27:233–255.

Korenman, S., and D. Neumark. 1991. Does marriage really make men more productive? *Journal of Human Resources* 26:282–307.

Korn, J. B. 1993. The fungible woman and other myths of sexual harassment. *Tulane Law Review* 67:1363–1419.

Krantz, L. 1992. *The Jobs Rated Almanac*. New York: World Almanac.

La Freniere, P., F. F. Strayer, and R. Gauthier. 1984. The emergence of same-sex affiliative preferences among preschool peers: A developmental/ethological perspective. *Child Development* 55:1958–1965.

Lacreuse, A., J. G. Herndon, R. J. Killiany, D. L. Rosene, and M. B. Moss. 1999. Spatial cognition in rhesus monkeys: Male superiority declines with age. *Hormones and Behavior* 36:70–76.

Lacy, W. B., J. L. Bokemeier, and J. M. Shepard. 1983. Job attribute preferences and work commitment of men and women in the United States. *Personnel Psychology* 36:315–329.

Lafontaine, E., and L. Tredeau. 1986. The frequency, sources, and correlates of sexual harassment among women in traditional male occupations. *Sex Roles* 15:433–442.

Lal, B., S. Yoon, and K. Carlson. 1999. *How Large Is the Gap in Salaries of Male and Female Engineers?* National Science Foundation, Division of Science Resources Studies Issue Brief. <http://www.nsf.gov/sbe/srs/issuebrf/sib99352.pdf>.

Lamb, M. E., M. A. Easterbrooks, and G. W. Holden. 1980. Reinforcement and punishment among preschoolers: Characteristics, effects, and correlates. *Child Development* 51:1230–1236.

Laumann, E. O., J. H. Gagnon, R. T. Michael, and S. Michaels. 1994. *The Social Organization of Sexuality: Sexual Practices in the United States.* Chicago: University of Chicago Press.

Le Boeuf, B. J. 1974. Male-male competition and reproductive success in elephant seals. *American Zoologist* 14:163–176.

Leibenluft, E., T. H. Dial, M. G. Haviland, and H. A. Pincus. 1993. Sex differences in rank attainment and research activities among academic psychiatrists. *Archives of General Psychiatry* 50:896–904.

Leinbach, M. D., and B. I. Fagot. 1993. Categorical habituation to male and female faces: Gender schematic processing in infancy. *Infant Behavior and Development* 16:317–332.

Lettau, M. 1994. *Compensation in Part-Time Jobs versus Full-Time Jobs: What If the Job Is the Same?* Bureau of Labor Statistics Working Paper, no. 260.

Lever, J. 1976. Sex differences in the games children play. *Social Problems* 23:478–487.

Leveroni, C. L., and S. A. Berenbaum. 1998. Early androgen effects on interest in infants: Evidence from children with congenital adrenal hyperplasia. *Developmental Neuropsychology* 14:321–340.

Levine, B. 1990. They face burgeoning demands at work and at home. *Los Angeles Times,* September 23, E–1.

Levine, S. C., J. Huttenlocher, A. Taylor, and A. Langrock. 1999. Early sex differences in spatial skill. *Developmental Psychology* 35:940–949.

Lewontin, R. C., S. Rose, and L. Kamin. 1984. *Not in Our Genes: Biology, Ideology, and Human Nature.* New York: Pantheon.

Liben, L. S. 1995. Psychology meets geography: Exploring the gender gap on the National Geography Bee. *Psychological Science Agenda* (January/February), 8–9.

Lillydahl, J. H. 1986. Women and traditionally male blue-collar jobs. *Work and Occupations* 13:307–323.

Lim, G. Y., and M. E. Roloff. 1999. Attributing sexual consent. *Journal of Applied Communication Research* 27:1–23.

Lindgren, J., and D. Seltzer. 1996. The most prolific law professors and faculties. *Chicago-Kent Law Review* 71:781–807.

Littleton, C. A. 1987. Reconstructing sexual equality. *California Law Review* 75:1279–1337.

Long, J. E. 1995. The effects of tastes and motivation on individual income. *Industrial and Labor Relations Review* 48:338–351.

Long, J. S. 1992. Measures of sex differences in scientific productivity. *Social Forces* 71:159–178.

Lorber, J. 1993. Believing is seeing: Biology as ideology. *Gender and Society* 7:568–581.

Lord, R. G., C. L. de Vader, and G. M. Alliger. 1986. A meta-analysis of the relation between personality traits and leadership perceptions: An application of validity generalization procedures. *Journal of Applied Psychology* 71:402–410.

Loscocco, K. A. 1990. Reactions to blue-collar work: A comparison of women and men. *Work and Occupations* 17:152–177.

Lott, B. 1996. Politics or science? The question of gender sameness/difference. *American Psychologist* 51:155–156.

Low, B. S. 1989. Cross-cultural patterns in the training of children: An evolutionary perspective. *Journal of Comparative Psychology* 103:311–319.

Low, B. S. 1992. Sex, coalitions, and politics in preindustrial societies. *Politics and the Life Sciences* 11:63–80.

Low, R., and R. Over. 1993. Gender differences in solution of algebraic word problems containing irrelevant information. *Journal of Educational Psychology* 85:331–339.

Lubinski, D. 2000. Scientific and social significance of assessing individual differences: "Sinking shafts at a few critical points." *Annual Review of Psychology* 51:405–444.

Lubinski, D., and C. P. Benbow. 1992. Gender differences in abilities and preferences among the gifted: Implications for the math-science pipeline. *Current Directions in Psychological Science* 1:61–66.

Lubinski, D., C. P. Benbow, and C. E. Sanders. 1993. Reconceptualizing gender differences in achievement among the gifted. In K. A. Heller, F. J. Moenks, and A. H. Passow (eds.), *International Handbook of Research and Development of Giftedness and Talent.* Elmsford, N.Y.: Pergamon Press.

Lueptow, L. B., L. Garovich, and M. B. Lueptow. 1995. The persistence of gender stereotypes in the face of changing sex roles: Evidence contrary to the sociocultural model. *Ethology and Sociobiology* 16:509–530.

Lummis, M., and H. W. Stevenson. 1990. Gender differences in beliefs and achievement: A cross-cultural study. *Developmental Psychology* 26:254–263.

Lundberg, U. 1983. Sex differences in behaviour pattern and catecholamine and cortisol excretion in 3–6-year-old day-care children. *Biological Psychology* 16:109–117.

Lundgren, D. C., E. B. Sampson, and M. B. Cahoon. 1998. Undergraduate men's and women's responses to positive and negative feedback about academic performance. *Psychological Reports* 82:87–93.

Lutner, R. E. 1993. Employer liability for sexual harassment: The morass of agency principles and respondeat superior. *University of Illinois Law Review* 1993:589–628.

Lynch, M., and K. Post. 1996. What glass ceiling? *Public Interest* 124:27–36.

Lyness, K. S., and D. E. Thompson. 1997. Above the glass ceiling? A comparison of matched samples of female and male executives. *Journal of Applied Psychology* 82:359–375.

Lynn, R. 1993. Sex differences in competitiveness and the valuation of money in twenty countries. *Journal of Social Psychology* 133:507–511.

Lyons, M. R. 1997. *Part-Time Work: Not a Problem Requiring a Solution.* Washington, D.C.: Employment Policy Foundation.

Lytton, H., and D. M. Romney. 1991. Parents' differential socialization of boys and girls: A meta-analysis. *Psychological Bulletin* 109:267–296.

MIT Committee on Women Faculty in the School of Science. 1999. *A Study on the Status of Women Faculty in Science at MIT.* <http://web.mit.edu/fnl/women/women.pdf>.

Maccoby, E. E. 1998. *The Two Sexes: Growing Up Apart, Coming Together.* Cambridge, Mass.: Belknap Harvard.

Maccoby, E. E. 1990. Gender and relationships: A developmental account. *American Psychologist* 45:513–520.

Maccoby, E. E. 1987. The varied meanings of "masculine" and "feminine." In J. M. Reinisch, L. A. Rosenblum, and S. A. Sanders (eds.), *Masculinity/Femininity: Basic Perspectives.* New York: Oxford University Press.

Maccoby, E. E., and C. N. Jacklin. 1974. *The Psychology of Sex Differences.* Stanford, Calif.: Stanford University Press.

Maccoby, E. E., and J. A. Martin. 1983. Socialization in the context of the family: Parent-child interaction. In E. M. Hetherington (ed.), *Handbook of Child Psychology,* vol. 4. New York: Wiley.

Maccoby, M. 1976. *The Gamesman: The New Corporate Leaders.* New York: Simon & Schuster.

MacCrimmon, K. R., and D. A. Wehrung. 1990. Characteristics of risk taking executives. *Management Science* 36:422–435.

MacKinnon, C. A. 1990. Legal perspectives on sexual difference. In D. L. Rhode (ed.), *Theoretical Perspectives on Sexual Difference.* New Haven, Conn.: Yale University Press.

Mahony, R. 1995. *Kidding Ourselves: Breadwinning, Babies, and Bargaining Power.* New York: Basic.

Major, B., and E. Konar. 1984. An investigation of sex differences in pay expectations and their possible causes. *Academy of Management Journal* 27:777–792.

Major, B., V. Vanderslice, and D. B. McFarlin. 1985. Effects of pay expected on pay received: The confirmatory nature of initial expectations. *Journal of Applied Social Psychology* 14:399–412.

Malin, M. H. 1998. Fathers and parental leave revisited. *Northern Illinois University Law Review* 19:25–56.

Malinowski, J. C., and W. T. Gillespie. 2001. Individual differences in performance on a large-scale, real-world wayfinding task. *Journal of Environmental Psychology* 21:73–82.

Mann, V. A., S. Sasanuma, N. Sakuma, and S. Masaki. 1990. Sex differences in cognitive abilities: A cross-cultural perspective. *Neuropsychologia* 28:1063–1077.

Mansnerus, L. 1993. Why women are leaving the law. *Working Woman,* April, 64–104.

Marcusson, H., and W. Oehmisch. 1977. Accident mortality in childhood in selected countries of different continents, 1950–1971. *World Health Statistical Report* 30:57.

Margolis, A. J., S. Greenwood, and D. Heilbron. 1983. Survey of men and women residents entering United States obstetrics and gynecology programs in 1981. *American Journal of Obstetrics and Gynecology* 146:541–546.

Markham, S. E., and G. H. McKee. 1995. Group absence behavior and standards: A multilevel analysis. *Academy of Management Journal* 38:1174–1190.

Marks, I. M. 1987. *Fears, Phobias, and Rituals: Panic, Anxiety, and Their Disorders.* New York: Oxford University Press.

Marshall, S. P., and J. D. Smith. 1987. Sex differences in learning mathematics: A longitudinal study with item and error analyses. *Journal of Educational Psychology* 79:372–383.

Martin, T., and B. Kirkcaldy. 1998. Gender differences on the EPQ-R and attitudes to work. *Personality and Individual Differences* 24:1–5.

Masica, D. N., J. Money, and A. A. Ehrhardt. 1971. Fetal feminization and female gender identity in the testicular feminizing syndrome of androgen insensitivity. *Archives of Sexual Behavior* 1:131–142.

Masters, M. S., and B. Sanders. 1993. Is the gender difference in mental rotation disappearing? *Behavior Genetics* 23:337–341.

Matteson, D. R. 1991. Attempting to change sex role attitudes in adolescents: Explorations of reverse effects. *Adolescence* 26:885–898.

Mazur, A., C. Halpern, and J. R. Udry. 1994. Dominant looking male teenagers copulate earlier. *Ethology and Sociobiology* 15:87–94.

Mazur, A., and A. Booth. 1998. Testosterone and dominance in men. *Behavioral and Brain Sciences* 21:353–397.

McBurney, D. H., S.J.C. Gaulin, T. Devineni, and C. Adams. 1997. Superior spatial memory of women: Stronger evidence for the gathering hypothesis. *Evolution and Human Behavior* 18:165–174.

McCauley, C. R. 1995. Are stereotypes exaggerated? A sampling of racial, gender, academic, occupational, and political stereotypes. In Y.-T. Lee, L. J. Jussim, and C. R. McCauley (eds.), *Stereotype Accuracy: Toward Appreciating Group Differences.* Washington, D.C.: American Psychological Association.

McClelland, D. C., and R. I. Watson. 1973. Power motivation and risk-taking behavior. *Journal of Personality* 41:121–139.

McCreary, D. R. 1994. The male role and avoiding femininity. *Sex Roles* 31:517–531.

McElrath, K. 1992. Gender, career disruption, and academic rewards. *Journal of Higher Education* 63:269–281.

McGuinness, D. 1988. Socialization versus biology: Time to move on. *Behavioral and Brain Sciences* 11:203–204.

McGuinness, D. 1979. How schools discriminate against boys. *Human Nature,* February, 82–88.

McGuinness, D. 1990. Behavioral tempo in pre-school boys and girls. *Learning and Individual Differences* 2:315–325.

McGuinness, D., and J. Symonds. 1977. Sex differences in choice behaviour: The object-person dimension. *Perception* 6:691–694.

McSherry, M. 2000. New formula helps women in math, science: Special dorm offers support at U. of I. *Chicago Tribune,* February 28.

Mead, M. 1928. *Coming of Age in Samoa.* New York: Blue Ribbon Books.

Mead, M. 1935. *Sex and Temperament in Three Primitive Societies.* New York: Wm. Morrow.

Mealey, L. 1992. Alternative adaptive models of rape. *Behavioral and Brain Sciences* 15:397–398.

Mealey, L. 2000. *Sex Differences: Developmental and Evolutionary Strategies.* San Diego, Calif.: Academic Press.

Meara, N. M., and J. D. Day. 1993. Perspectives on achieving via interpersonal competition between college men and college women. *Sex Roles* 28:91–110.

Megargee, E. I. 1969. Influence of sex roles on the manifestation of leadership. *Journal of Applied Psychology* 53:377–382.

Merritt, D. J., B. F. Reskin, and M. Fondell. 1993. Family, place, and career: The gender paradox in law school hiring. *Wisconsin Law Review* 1993:395–463.

Miles, C., R. Green, G. Sanders, and M. Hines. 1998. Estrogen and memory in a transsexual population. *Hormones and Behavior* 34:199–208.

Miller, A. A. 1985. A developmental study of the cognitive basis of performance impairment after failure. *Journal of Personality and Social Psychology* 49:529–538.

Miller, G. F. 2000. *The Mating Mind: How Sexual Choice Shaped the Evolution of Human Nature.* New York: Doubleday.

Mincer, J., and H. Ofek. 1982. Interrupted work careers: Depreciation and restoration of human capital. *Journal of Human Resources* 17:3–24.

Minor, L. L., and C. P. Benbow. 1996. Construct validity of the SAT-M: A comparative study of high school students and gifted seventh graders. In C. P. Benbow and D. Lubinski (eds.), *Intellectual Talent: Psychometric and Social Issues.* Baltimore, Md.: Johns Hopkins University Press.

Mirowsky, J., and C. E. Ross. 1995. Sex differences in distress: Real or artifact? *American Sociological Review* 60:449–468.

Moffat, S. D., and E. Hampson. 1996. A curvilinear relationship between testosterone and spatial cognition in humans: Possible influence of hand preference. *Psychoneuroendocrinology* 21:323–337.

Moffat, S. D., E. Hampson, and M. Hatzipantelis. 1998. Navigation in a "virtual" maze: Sex differences and correlation with psychometric measures of spatial ability in humans. *Evolution and Human Behavior* 19:73–87.

Moir, A., and D. Jessel. 1989. *Brain Sex: The Real Difference between Men and Women.* New York: Dell.

Morrison, A. M., R. P. White, E. Van Velsor, and the Center for Creative Leadership. 1992. *Breaking the Glass Ceiling: Can Women Reach the Top of America's Largest Corporations?* (Updated ed.). Reading, Mass.: Addison-Wesley.

Morrongiello, B. A., and T. Dawber. 1999. Parental influences on toddlers' injury-risk behaviors: Are sons and daughters socialized differently? *Journal of Applied Developmental Psychology* 20:227–251.

Morrongiello, B. A., and T. Dawber. 1998. Toddlers' and mothers' behaviors in an injury-risk situation: Implications for sex differences in childhood injuries. *Journal of Applied Developmental Psychology* 19:625–639.

Morrongiello, B. A., and H. Rennie. 1998. Why do boys engage in more risk taking than girls? The role of attributions, beliefs, and risk appraisals. *Journal of Pediatric Psychology* 23:33–43.

Moskal, B. S. 1997. Women make better managers. *Industry Week,* February 3.

Muehlenhard, C. L., and L. C. Hollabaugh. 1988. Do women sometimes say no when they mean yes? The prevalence and correlates of women's token resistance to sex. *Journal of Personality and Social Psychology* 54:872–879.

Muehlenhard, C. L., and M. L. McCoy. 1991. Double standard/double bind: The sexual double standard and women's communication about sex. *Psychology of Women Quarterly* 15:447–461.

Murray, M. A., and T. Atkinson. 1981. Gender differences in correlates of job satisfaction. *Canadian Journal of Behavioural Science* 13:44–52.

Nakosteen, R. A., and M. A. Zimmer. 1987. Marital status and earnings of young men: A model with endogenous selection. *Journal of Human Resources* 22:248–268.

National Center for Education Statistics. 2000. *Trends in Educational Equity of Girls and Women.* U.S. Department of Education, Office of Educational Research and Improvement, NCES 2000–030. <http://nces.ed.gov/pubs2000/2000030.pdf>.

National Center for Education Statistics. 1998. *Pursuing Excellence: A Study of U.S. Twelfth-Grade Mathematics and Science Achievement in International Context: Initial Findings from the Third International Mathematics and Science Study.* U.S. Department of Education, Office of Educational Research and Improvement, NCES 98–049. <http://nces.ed.gov/pubs98/98049.pdf>.

National Science Foundation. 2000. *Women, Minorities, and Persons with Disabilities in Science and Engineering, 2000.* Arlington, VA (NSF 00–327). <http://www.nsf.gov/sbe/srs/nsf00327/pdf/document.pdf>.

National Science Foundation. 1999. *Women, Minorities, and Persons with Disabilities in Science and Engineering, 1998.* Arlington, VA (NSF 99–338). <http://www.nsf.gov/sbe/srs/nsf99338/pdf/document.pdf>.

National Science Foundation. 1996. *Women, Minorities, and Persons with Disabilities in Science and Engineering, 1996.* Arlington, VA (NSF 96–311) <http://www.nsf.gov/sbe/srs/nsf96311/intro.pdf>.

Nelson, B. A., E. M. Opton Jr., and T. E. Wilson. 1980. Wage discrimination and the "comparable worth" theory in perspective. *University of Michigan Journal of Law Reform* 13:233–301.

New York Times. 1999. Clinton seeks more money to reduce gap in wages. January 31.

New York Times. 1999. Editorial: Gender bias on the campus. March 28.

New York Times. 1993. Editorial: High murder rate for women on job. October 3.

Nicholson, N. 1998. How hardwired is human behavior? *Harvard Business Review* 76(4): 134–147.

Nordic Business Report. 2000. Parental leave in Sweden causes mixed feelings, June 29.

Nordvik, H., and B. Amponsah. 1998. Gender differences in spatial abilities and spatial activity among university students in an egalitarian educational system. *Sex Roles* 38:1009–1023.

Nossel, S., and E. Westfall. 1998. *Presumed Equal: What America's Top Women Lawyers Really Think about Their Firms*. Franklin Lakes, N.J.: Career Press.

Nyborg, H. 1994. *Hormones, Sex, and Society: The Science of Physicology*. Westport, Conn.: Praeger.

O'Brien, M., and A. C. Huston. 1985. Development of sex-typed play behavior in toddlers. *Developmental Psychology* 21:866–871.

O'Farrell, B. 1999. Women in blue-collar and related occupations at the end of the millennium. *Quarterly Review of Economics and Finance* 39:699–722.

O'Neill, J. 1990. Women and wages. *American Enterprise*, November/December, 25–33.

O'Neill, J., and S. Polachek. 1993. Why the gender gap in wages narrowed in the 1980s. *Journal of Labor Economics* 11:205–228.

Okin, S. M. 1989. *Justice, Gender, and the Family*. New York: Basic.

Oldham, C. 2000. More employees squeezing shopping into workdays: Firms take into account time crush during holidays. *Dallas Morning News*, December 17.

Olian, J. D., D. P. Schwab, and Y. Haberfeld. 1988. The impact of applicant gender compared to qualifications on hiring recommendations: A meta-analysis of experimental studies. *Organizational Behavior and Human Decision Processes* 41:180–195.

Olsen, R. A., and C. M. Cox. 2001. The influence of gender on the perception and response to investment risk: The case of professional investors. *Journal of Psychology and Financial Markets* 2:29–36.

Olweus, D., A. Mattsson, D. Schalling, and H. Loew. 1980. Testosterone, aggression, physical, and personality dimensions in normal adolescent males. *Psychosomatic Medicine* 42:253–269.

Omark, D. R., and M. S. Edelman. 1975. A comparison of status hierarchies in young children: An ethological approach. *Social Science Information* 14(5): 87–107.

Oshige, M. 1995. What's sex got to do with it? *Stanford Law Review* 47:565–594.

Packer, C., D. A. Collins, A. Sindimwo, and J. Goodall. 1995. Reproductive constraints on aggressive competition in female baboons. *Nature* 373:60–63.

Padavic, I. 1992. White-collar work values and women's interest in blue-collar jobs. *Gender and Society* 6:215–230.

Padavic, I. 1991. Attractions of male blue-collar jobs for black and white women: Economic need, exposure, and attitudes. *Social Science Quarterly* 72:33–49.

Padavic, I., and B. F. Reskin. 1990. Men's behavior and women's interest in blue-collar jobs. *Social Problems* 37:613–628.

Paetzold, R. L., and S. L. Willborn. 1994. *The Statistics of Discrimination: Using Statistical Evidence in Discrimination Cases*. Colorado Springs, Colo.: Shepard's/McGraw-Hill.

Paglin, M., and A. M. Rufolo. 1990. Heterogeneous human capital, occupational choice, and male-female earnings differences. *Journal of Labor Economics* 8: 123–144.

Palepu, A., R. H. Friedman, R. C. Barnett, P. L. Carr, A. S. Ash, L. Szalacha, and M. A. Moskowitz. 1998. Junior faculty members' mentoring relationships and their professional development in U.S. medical schools. *Academic Medicine* 73:318–323.

Palmer, D. L., and T. Folds-Bennett. 1998. Performance on two attention tasks as a function of sex and competition. *Perceptual and Motor Skills* 86: 363–370.

Parasuraman, S., Y. S. Purohit, V. M. Godshalk, and N. J. Beutell. 1996. Work and family variables, entrepreneurial career success, and psychological well-being. *Journal of Vocational Behavior* 48: 275–300.

Patai, D. 1998. *Heterophobia: Sexual Harassment and the Future of Feminism.* Lanham, Md.: Rowman & Littlefield.

Patterson, A. S. 2000. Women in global politics: Progress or stagnation? *USA Today* (Magazine), September 1.

Paul, E. 1989. *Equity and Gender: The Comparable Worth Debate.* New Brunswick, N.J.: Transaction.

Paul, E. F. 1990. Sexual harassment as sex discrimination: A defective paradigm. *Yale Law and Policy Review* 8: 333–365.

Pedersen, F. A., and R. Q. Bell. 1970. Sex differences in preschool children without histories of complications of pregnancy and delivery. *Developmental Psychology* 3: 10–15.

Pettit, G. S., A. Bakshi, K. A. Dodge, and J. D. Coie. 1990. The emergence of social dominance in young boys' play groups: Developmental differences and behavior correlates. *Developmental Psychology* 26: 1017–1025.

Pheasant, S. T. 1983. Sex differences in strength: Some observations on their variability. *Applied Ergonomics* 14: 205–211.

Phillips-Miller, D., C. Morrison, and N. J. Campbell. 2001. Same profession, different career: A study of men and women in veterinary medicine. In F. Columbus (ed.), *Advances in Psychology Research,* vol. 5. Huntington, N.Y.: Nova Science.

Plavcan, J. M., and C. P. Van Schaik. 1997. Intrasexual competition and body weight dimorphism in anthropoid primates. *American Journal of Physical Anthropology* 103:37–68.

Pleck, J. H. 1997. Paternal involvement: Levels, sources, and consequences. In M. E. Lamb (ed.), *The Role of the Father in Child Development.* 3rd ed. New York: Wiley.

Plomin, R. 1990. *Nature and Nurture: An Introduction to Human Behavioral Genetics.* Pacific Grove, Calif.: Brooks/Cole.

Polachek, S. W. 1987. Occupational segregation and the gender wage gap. *Population Research and Policy Review* 6: 47–67.

Polachek, S. W. 1995. Human capital and the gender earnings gap: A response to feminist critiques. In E. Kuiper and J. Sap (eds.), *Out of the Margin: Feminist Perspectives on Economics.* New York: Routledge.

Popenoe, D. 1996. *Life without Father.* New York: Free Press.

Popenoe, D. 1991. Family decline in the Swedish welfare state. *The Public Interest* 102(winter): 65–77.

Powell, G. N., and D. A. Butterfield. 1994. Investigating the "glass ceiling" phenomenon: An empirical study of actual promotions to top management. *Academy of Management Journal* 37: 68–86.

Press, J. E., and E. Townsley. 1998. Wives' and husbands' housework reporting: Gender, class and social desirability. *Gender and Society* 12: 188–218.

Preston, K., and K. Stanley. 1987. "What's the worst thing . . . ?" Gender-directed insults. *Sex Roles* 17: 209–219.

Primack, R. B., and V. E. O'Leary. 1989. Research productivity of men and women ecologists: A longitudinal study of former graduate students. *Bulletin of the Ecological Society of America* 70:7–12.

Primack, R. B., and V. E. O'Leary. 1993. Cumulative disadvantages in the careers of women ecologists. *BioScience* 43:158–165.

Pryor, J. B., and J. D. Day. 1988. Interpretations of sexual harassment: An attributional analysis. *Sex Roles* 18:405–417.

Pryor, J. B., C. M. LaVite, and L. M. Stoller. 1993. A social psychological analysis of sexual harassment: The person/situation interaction. *Journal of Vocational Behavior* 42:68–83.

Purifoy, F. E., and L. H. Koopmans. 1979. Androstenedione, testosterone, and free testosterone concentration in women of various occupations. *Social Biology* 26:179–188.

Quadagno, D. M., R. Briscoe, and J. S. Quadagno. 1977. Effect of perinatal gonadal hormones on selected nonsexual behavior patterns: A critical assessment of the nonhuman and human literature. *Psychological Bulletin* 84:62–80.

Quinn, R. E. 1977. Coping with Cupid: The formation, impact, and management of romantic relationships in organizations. *Administrative Science Quarterly* 22: 30–45.

Ragins, B. R., and T. A. Scandura. 1995. Antecedents and work-related correlates of reported sexual harassment: An empirical investigation of competing hypotheses. *Sex Roles* 32:429-455.

Raymond, C. L., and C. P. Benbow. 1986. Gender differences in mathematics: A function of parental support and student sex typing? *Developmental Psychology* 22: 808–819.

Redman, S., D. Saltman, J. Straton, B. Young, and C. Paul. 1994. Determinants of career choices among women and men medical students and interns. *Medical Education* 28:361–371.

Reinisch, J. M., M. Ziemba-Davis, and S. A. Sanders. 1991. Hormonal contributions to sexually dimorphic behavioral development in humans. *Psychoneuroendocrinology* 16:213–278.

Reis, S. B., I. P. Young, and J. C. Jury. 1999. Female administrators: A crack in the glass ceiling. *Journal of Personnel Evaluation in Education* 13:71–82.

Reiser, L. W., W. H. Sledge, W. Fenton, and P. Leaf. 1993. Beginning careers in academic psychiatry for women—"Bermuda triangle"? *American Journal of Psychiatry* 150:1392–1397.

Reskin, B. F., and P. A. Roos. 1990. *Job Queues, Gender Queues: Explaining Women's Inroads into Male Occupations*. Philadelphia: Temple University Press.

Rhoads, S. E. 1993. *Incomparable Worth: Pay Equity Meets the Market.* Cambridge: Cambridge University Press.

Rhode, D. L. 1991. The "no-problem" problem: Feminist challenges and cultural change. *Yale Law Journal* 100:1731–1793.

Rice, W. R. 1996. Sexually antagonistic male adaptation triggered by experimental arrest of female evolution. *Nature* 381:232–234.

Roback, J. 1993. Beyond equality. *Georgetown Law Journal* 82:121–133.

Robinson, J. P., and A. Bostrom. 1994. The overestimated workweek? What time diary measures suggest. *Monthly Labor Review* 117(8): 11–23.

Robinson, J. P., and G. Godbey. 1999. *Time for Life: The Surprising Ways Americans Use Their Time.* 2nd ed. University Park: Pennsylvania State University Press.

Robinson, N. M., R. D. Abbott, V. W. Berninger, and J. Busse. 1996. The structure of abilities in math-precocious young children: Gender similarities and differences. *Journal of Educational Psychology* 88:341–352.

Rosenthal, R., and D. B. Rubin. 1982. Further meta-analytic procedures for assessing cognitive gender differences. *Journal of Educational Psychology* 74:708–712.

Rossi, A. S. 1977. A biosocial perspective on parenting. *Daedalus* 106(2): 1–31.

Roth, P. 1969. *Portnoy's Complaint.* New York: Random House.

Rotundo, M., D.-H. Nguyen, and P. R. Sackett. 2001. A meta-analytic review of gender differences in perceptions of sexual harassment. *Journal of Applied Psychology* 86:914–922.

Ruhm, C. 1998. The economic consequences of parental leave mandates: Lessons from Europe. *Quarterly Journal of Economics* 113:285–317.

Rushton, J. P., D. W. Fulker, M. C. Neale, D. K. Nias, and H. J. Eysenck. 1986. Altruism and aggression: The heritability of individual differences. *Journal of Personality and Social Psychology* 50:1192–1198.

Sadalla, E. K., D. T. Kenrick, and B. Vershure. 1987. Dominance and heterosexual attraction. *Journal of Personality and Social Psychology* 52:730–738.

Sandberg, D. E., and H.F.L. Meyer-Bahlburg. 1994. Variability in middle childhood play behavior: Effects of gender, age, and family background. *Archives of Sexual Behavior* 23:645–663.

Savin-Williams, R. C. 1979. Dominance hierarchies in groups of early adolescents. *Child Development* 50:923–935.

Scales, A. 1981. Towards a feminist jurisprudence. *Indiana Law Journal* 56:375–444.

Scarr, S., and K. McCartney. 1983. How people make their own environments: A theory of genotype—environment effects. *Child Development* 54:424–435.

Schiavi, R. C., A. Theilgaard, D. R. Owen, and D. White. 1984. Sex chromosome anomalies, hormones, and aggressivity. *Archives of General Psychiatry* 41:93–99.

Schneider, B. E. 1984. The office affair: Myth and reality for heterosexual and lesbian women workers. *Sociological Perspectives* 27:443– 464.

Schneider, B. E. 1982. Consciousness about sexual harassment among heterosexual and lesbian women workers. *Journal of Social Issues* 38(4): 75–97.

Schooley, R. L., C. R. McLaughlin, G. J. Matula Jr., and W. B. Krohn. 1994. Denning chronology of female black bears: Effects of food, weather, and reproduction. *Journal of Mammalogy* 75:466–477.

Schor, J. B. 1994. Worktime in contemporary context: Amending the Fair Labor Standards Act. *Chicago-Kent Law Review* 70:157–172.

Schor, J. B. 1991. *The Overworked American: The Unexpected Decline of Leisure.* New York: Basic.

Schrader, M. P., and D. L. Wann. 1999. High-risk recreation: The relationship between participant characteristics and degree of involvement. *Journal of Sport Behavior* 22:426–441.

Schultz, V. 1990. Telling stories about women and work: Judicial interpretations of sex segregation in the workplace in Title VII cases raising the lack of interest argument. *Harvard Law Review* 103:1749–1843.

Schultz, V., and S. Petterson. 1992. Race, gender, work, and choice: An empirical study of the lack of interest defense in Title VII cases challenging job segregation. *University of Chicago Law Review* 59:1073–1181.

Schwab, S. J. 1993. Life-cycle justice: Accommodating just cause and employment at will. *Michigan Law Review* 92:8–62.

Schwartz, F. N. 1989. Management women and the new facts of life. *Harvard Business Review* 67(January–February): 65–76.

Schwartz, F. N. 1992. *Breaking with Tradition: Women and Work, The New Facts of Life.* New York: Warner.

Scott, K. D., and E. L. McLellan. 1990. Gender differences in absenteeism. *Public Personnel Management* 19:229–253.

Selmi, M. 2000. Family leave and the gender wage gap. *North Carolina Law Review* 78:707–782.

Semonsky, M. R., and L. V. Rosenfeld. 1994. Perceptions of sexual violations: Denying a kiss, stealing a kiss. *Sex Roles* 30:503–520.

Servin, A., G. Bohlin, and L. Berlin. 1999. Sex differences in 1-, 3-, and 5-year-olds' toy-choice in a structured play-session. *Scandinavian Journal of Psychology* 40: 43–48.

Sexton, D. L., and N. Bowman-Upton. 1990. Female and male entrepreneurs: Psychological characteristics and their role in gender-related discrimination. *Journal of Business Venturing* 5:29–36.

Seymour, E., and N. M. Hewitt. 1997. *Talking about Leaving: Why Undergraduates Leave the Sciences.* Boulder, Colo.: Westview Press.

Shauman, K. A., and Y. Xie. 1996. Geographic mobility of scientists: Sex differences and family constraints. *Demography* 33:455–468.

Shaw, L. B., and D. Shapiro. 1987. Women's work plans: Contrasting expectations and actual work experience. *Monthly Labor Review* 110 (11): 7–13.

Shelton, B. A. 1992. *Women, Men and Time: Gender Differences in Paid Work, Housework and Leisure.* Westport, Conn.: Greenwood.

Shelton, B. A., and D. John. 1993. Does marital status make a difference? Housework among married and cohabiting men and women. *Journal of Family Issues* 14:401–420.

Shotland, R. L., and J. M. Craig. 1988. Can men and women differentiate between friendly and sexually interested behavior? *Social Psychology Quarterly* 51:66–73.

Shye, D. 1991. Gender differences in Israeli physicians' career patterns, productivity and family structure. *Social Science and Medicine* 32:1169–1181.

Silverman I., J. Choi, A. Mackewn, M. Fisher, J. Moro, and E. Olshansky. 2000. Evolved mechanisms underlying wayfinding: Further studies on the hunter-gatherer theory of spatial sex differences. *Evolution and Human Behavior* 21: 201–213.

Silverman, I., S. E. Dickens, M. Eals, and J. Fine. 1993. Perceptions of maternal and paternal solicitude by birth children and adoptees: Relevance for maternal bonding theory. *Early Development and Parenting* 2:135–144.

Silverman, I., and M. Eals. 1992. Sex differences in spatial abilities: Evolutionary theory and data. In J. H. Barkow, L. Cosmides, and J. Tooby (eds.), *The Adapted Mind: Evolutionary Psychology and the Generation of Culture.* New York: Oxford University Press.

Silverman, I., K. Phillips, and L. K. Silverman. 1996. Homogeneity of effect sizes for sex across spatial tests and cultures: Implications for hormonal theories. *Brain and Cognition* 31:90–94.

Simmons, W. W. 2001. When it comes to choosing a boss, Americans still prefer men: Even women prefer a male boss by two-to-one margin. *Gallup News Service,* January 11. <http://www.gallup.com/poll/releases/pr010111.asp>.

Slabbekoorn, D., S.H.M. van Goozen, J. Megens, L.J.G. Gooren, and P. T. Cohen-Kettenis. 1999. Activating effects of cross-sex hormones on cognitive functioning: A study of short-term and long-term hormone effects in transsexuals. *Psychoneuroendocrinology* 24:423–447.

Smith, A. 1976 [1776]. *An Inquiry into the Nature and Causes of the Wealth of Nations.* New York: Oxford University Press.

Smith, C. J., P. Rodenhauser, and R. J. Markert. 1991. Gender bias of Ohio physicians in the evaluation of the personal statements of residency applicants. *Academic Medicine* 66:479–481.

Smuts, B. 1987. Gender, aggression, and influence. In B. B. Smuts, D. L. Cheney, R. M. Seyfarth, R. W. Wrangham, and T. T. Struhsaker (eds.), *Primate Societies.* Chicago: University of Chicago Press.

Smuts, B. 1995. The evolutionary origins of patriarchy. *Human Nature* 6:1–32.

Solberg, E., and T. Laughlin. 1995. The gender pay gap, fringe benefits, and occupational crowding. *Industrial and Labor Relations Review* 48:692–708.

Sommers, C. H. 2000. *The War against Boys: How Misguided Feminism Is Harming Our Young Men.* New York: Simon & Schuster.

South, S. J., and G. Spitze. 1994. Housework in marital and nonmarital households. *American Sociological Review* 59:327–347.

Spiro, M. E. 1996. *Gender and Culture: Kibbutz Women Revisited.* New Brunswick, N.J.: Transaction.

Staines, G. L., K. J. Pottick, and D. A. Fudge. 1986. Wives' employment and husbands' attitudes toward work and life. *Journal of Applied Psychology* 71:118–128.

Stanley, J. C., and H. Stumpf. 1996. Able youths and achievement tests. *Behavioral and Brain Sciences* 19:263–264.

Stanley, S. C., J. G. Hunt, and L. L. Hunt. 1986. The relative deprivation of husbands in dual-earner households. *Journal of Family Issues* 7:3–20.

Stockdale, M. S. 1993. The role of sexual misperceptions of women's friendliness in an emerging theory of sexual harassment. *Journal of Vocational Behavior* 42:84–101.

Storey, A. E., C. J. Walsh, R. L. Quinton, and K. E. Wynne-Edwards. 2000. Hormonal correlates of paternal responsiveness in new and expectant fathers. *Evolution & Human Behavior* 21:79–95.

Strenta, C., R. Elliott, R. Adair, M. Matier, and J. Scott. 1994. Choosing and leaving science in highly selective institutions. *Research in Higher Education* 35:513–547.

Struckman-Johnson, C., and D. Struckman-Johnson. 1994. Men's reactions to hypothetical female sexual advances: A beauty bias in response to sexual coercion. *Sex Roles* 31:387–405.

Struckman-Johnson, C., and D. Struckman-Johnson. 1993. College men's and women's reactions to hypothetical sexual touch varied by initiator gender and coercion level. *Sex Roles* 29:371–385.

Studd, M. V., and U. E. Gattiker. 1991. The evolutionary psychology of sexual harassment in organizations. *Ethology and Sociobiology* 12:249–290.

Stuhlmacher, A. F., and A. E. Walters. 1999. Gender differences in negotiation outcome: A meta-analysis. *Personnel Psychology* 52:653–677.

Subich, L. M., G. V. Barrett, D. Doverspike, and R. A. Alexander. 1989. The effects of sex-role-related factors on occupational choice and salary. In R. T. Michael, H. I. Hartmann, and B. O'Farrell (eds.), *Pay Equity: Empirical Inquiries.* Washington, D.C.: National Academy Press.

Summers, L. H. 1989. Some simple economics of mandated benefits. *American Economic Review* 79:177–183.

Swim, J., E. Borgida, G. Maruyama, and D. G. Myers. 1989. Joan McKay versus John McKay: Do gender stereotypes bias evaluations? *Psychological Bulletin* 105:409–429.

Tam, T. 1997. Sex segregation and occupational gender inequality in the United States:

Devaluation or specialized training? *American Journal of Sociology* 102:1652–1692.

Tangri, S. S., M. R. Burt, and L. B. Johnson. 1982. Sexual harassment at work: Three explanatory models. *Journal of Social Issues* 38:33–54.

Tanner, J., R. Cockerill, J. Barnsley, and A. P. Williams. 1999. Gender and income in pharmacy: Human capital and gender stratification theories revisited. *British Journal of Sociology* 50:97–117.

Taylor, J. K. 1999. *What to Do When You Don't Want to Call the Cops: A Non-Adversarial Approach to Sexual Harassment.* New York: New York University Press.

Terpstra, D. E., and D. D. Baker. 1986. A framework for the study of sexual harassment. *Basic and Applied Social Psychology* 7:17–34.

Terpstra, D. E., and S. E. Cook. 1985. Complainant characteristics and reported behaviors and consequences associated with formal sexual harassment charges. *Personnel Psychology* 38:559–574.

Tesch, B. J., H. M. Wood, A. L. Helwig, and A. B. Nattinger. 1995. Promotion of women physicians in academic medicine: Glass ceiling or sticky floor? *Journal of the American Medical Association* 273:1022–1025.

Thornhill, N. W. 1996. Psychological adaptation to sexual coercion in victims and offenders. In D. M. Buss and N. M. Malamuth (eds.), *Sex, Power, Conflict: Evolutionary and Feminist Perspectives.* New York: Oxford University Press.

Tiger, L. 1997. Are the harassers in charge? Comment on article by Kingsley Browne. *Journal of Contemporary Legal Issues* 8:79–86.

Tiger, L. 1988. Sex differences in mathematics: Why the fuss? *Behavioral and Brain Sciences* 11:212.

Tiger, L. 1987. Alienated from the meanings of reproduction? In J. M. Reinisch, L. A. Rosenblum, and S. A. Sanders (eds.), *Masculinity/Femininity: Basic Perspectives.* New York: Oxford University Press.

Tiger, L. 1979. Biology, psychology, and incorrect assumptions of cultural relativism. In N. A. Chagnon and W. Irons (eds.), *Evolutionary Biology and Human Social Behavior.* North Scituate, Mass.: Duxbury.

Tiger, L., and J. Shepher. 1975. *Women in the Kibbutz.* New York: Harcourt Brace Jovanovich.

Tolbert, P. S., and P. Moen. 1998. Men's and women's definitions of "good" jobs: Similarities and differences by age and across time. *Work and Occupations* 25:168–194.

Tooby, J., and L. Cosmides. 1990. On the universality of human nature and the uniqueness of the individual: The role of genetics and adaptation. *Journal of Personality* 58:17–67.

Toscano, G. 1997. Dangerous jobs. In Bureau of Labor Statistics, *Compensation and Working Conditions,* summer, 57–60.

Toscano, G. A., and J. A. Windau. 1998. Profile of fatal work injuries in 1996. In Bureau of Labor Statistics, *Compensation and Working Conditions,* spring, 37–45.

Townsend, B. 1996. Room at the top for women. *American Demographics* 18(7): 28–33.

Townsend, J. M. 1999. Male dominance hierarchies and women's intrasexual competition. *Behavioral and Brain Sciences* 22:235–236.

Townsend, J. M. 1989. Mate selection criteria: A pilot study. *Ethology and Sociobiology* 10:241–253.

Townsend, J. M. 1987. Sex differences in sexuality among medical students: Effects of increasing socioeconomic status. *Archives of Sexual Behavior* 16:425–444.

Trankina, M. L. 1993. Gender differences in attitudes toward science. *Psychological Reports* 73:123–130.

Tremblay, R. E., B. Schaal, B. Boulerice, L. Arseneault, R. G. Soussignan, D. Paquette, and D. Laurent. 1998. Testosterone, physical aggression, dominance, and physical development in early adolescence. *International Journal of Behavioral Development* 22:753–777.

Trivers, R. L. 1972. Parental investment and sexual selection. In B. G. Campbell (ed.), *Sexual Selection and the Descent of Man.* Chicago: Aldine.

Trivers, R. L. 1985. *Social Evolution.* Menlo Park, Calif.: Benjamin/Cummings.

Udry, J. R. 2000. Biological limits of gender construction. *American Sociological Review* 65:443–457.

Udry, J. R., and B. K. Eckland. 1984. Benefits of being attractive: Differential payoffs for men and women. *Psychological Reports* 54:47–56.

Udry, J. R., N. M. Morris, and J. Kovenock. 1995. Androgen effects on women's gendered behaviour. *Journal of Biosocial Science* 27:359–368.

Valian, V. 1998. *Why So Slow? The Advancement of Women.* Cambridge, Mass.: MIT Press.

Valliere, R. 2000. Boston Fire Department lagging in efforts to include women, minorities, report says. *Daily Labor Report* 15:A–8.

Van Goozen, S.H.M., P. T. Cohen-Kettenis, L.J.G. Gooren, N. H. Frijda, and N. E. Van de Poll. 1995. Gender differences in behaviour: Activating effects of cross-sex hormones. *Psychoneuroendocrinology* 20:343–363.

Van Goozen, S.H.M., P. T. Cohen-Kettenis, L.J.G. Gooren, N. H. Frijda, and N. E. Van de Poll. 1994. Activating effects of androgens on cognitive performance: Causal evidence in a group of female-to-male transsexuals. *Neuropsychologia* 32:1153–1157.

Veevers, J. E., and E. M. Gee. 1986. Playing it safe: Accident mortality and gender roles. *Sociological Focus* 19:349–360.

Vitug, M. D. 1994. The Philippines: Fighting the patriarchy in growing numbers. *Science* 263:1491–1492.

Voyer, D., S. Voyer, and M. P. Bryden. 1995. Magnitude of sex differences in spatial abilities: A meta-analysis and consideration of critical variables. *Psychological Bulletin* 117:250–270.

Waal, F. de. 2001. *The Ape and the Sushi Master: Cultural Reflections of a Primatologist.* New York: Basic.

Waldfogel, J. 1998. The family gap for young women in the United States and Britain: Can maternity leave make a difference? *Journal of Labor Economics* 16:505–545.

Ward, O. B., A. M. Wexler, J. R. Carlucci, M. A. Eckert, and I. L. Ward. 1996. Critical periods of sensitivity of sexually dimorphic spinal nuclei to prenatal testosterone exposure in female rats. *Hormones and Behavior* 30:407–415.

Watson, J. B. 1925. *Behaviorism.* New York: W. W. Norton.

Weinberg, B. A. 2000. Computer use and the demand for female workers. *Industrial and Labor Relations Review* 53:290–308.

Weinberg, R. S., and J. Ragan. 1979. Effects of competition, success/failure, and sex on intrinsic motivation. *Research Quarterly* 50:503–510.

Weisfeld, C. C. 1986. Female behavior in mixed-sex competition: A review of the literature. *Developmental Review* 6:278–299.

Weisman, C. S., M. A. Teitelbaum, C. A. Nathanson, G. A. Chase, T. M. King, and D. M. Levine. 1986. Sex differences in the practice patterns of recently trained obstetrician-gynecologists. *Obstetrics and Gynecology* 67:776–782.

Wells, D. L., and P. G. Hepper. 1999. Male and female dogs respond differently to men and women. *Applied Animal Behaviour Science* 61:341–349.

White, D. G. 1985. *Ar'n't I a Woman? Female Slaves in the Plantation South.* New York: W. W. Norton.

Whiting, B. B., and C. P. Edwards. 1988. *Children of Different Worlds: The Formation of Social Behavior.* Cambridge, Mass.: Harvard University Press.

Whittington, L. A., J. Alm, and H. E. Peters. 1990. Fertility and the personal exemption: Implicit pronatalist policy in the United States. *American Economic Review* 80:545–556.

Wiener, R. L., B. A. Watts, K. H. Goldkamp, and C. Gasper. 1995. Social analytic investigation of hostile work environments: A test of the reasonable woman standard. *Law and Human Behavior* 19:263–281.

Wilberg, S., and R. Lynn. 1999. Sex differences in historical knowledge and school grades: A 26 nation study. *Personality and Individual Differences* 27:1221–1229.

Williams, C. 1992. The glass escalator: Hidden advantages for men in the "female" professions. *Social Problems* 39:253–267.

Williams, C. J. 2000. When love is never having to say "I do." *Los Angeles Times,* March 31, A–1.

Williams, C. L. 1989. *Gender Differences at Work: Women and Men in Nontraditional Occupations.* Berkeley: University of California Press.

Williams, C. L., and W. H. Meck. 1991. The organizational effects of gonadal steroids on sexually dimorphic spatial ability. *Psychoneuroendocrinology* 16:155–176.

Williams, J. 1991. Gender wars: Selfless women in the republic of choice. *New York University Law Review* 66:1559–1634.

Williams, J. 2000. *Unbending Gender: Why Family and Work Conflict and What to Do about It.* New York: Oxford University Press.

Williams, J. C. 1989. Deconstructing gender. *Michigan Law Review* 87:797–845.

Williams, J. E., and D. L. Best. 1990. *Measuring Sex Stereotypes: A Multination Study.* (Rev. ed.). Newbury Park, Calif.: Sage.

Willingham, W. W., and N. S. Cole. 1997. *Gender and Fair Assessment.* Mahwah, N.J.: Erlbaum.

Willingham, W. W., and L. M. Johnson. 1997. *Supplement to Gender and Fair Assessment.* Mahwah, N.J.: Erlbaum.

Wilson, M., and M. Daly. 1985. Competitiveness, risk taking, and violence: The young male syndrome. *Ethology and Sociobiology* 6:59–73.

Winter, P. A., and J. M. Butters. 1998. An investigation of dental student practice preferences. *Journal of Dental Education* 62:565–572.

Wong, P.T.P., G. E. Kettlewell, and C. F. Sproule. 1985. On the importance of being masculine: Sex role, attribution, and women's career achievement. *Sex Roles* 12:757–769.

Wood, R. G., M. E. Corcoran, and P. N. Courant. 1993. Pay differences among the highly paid: The male-female earnings gap in lawyers' salaries. *Journal of Labor Economics* 11:417–441.

Workman, J. E., and K. P. Johnson. 1991. The role of cosmetics in attributions about sexual harassment. *Sex Roles* 24:759–769.

Wulf, W. A. 1999. Testimony to the Commission on the Advancement of Women and Minorities in Science, Engineering, and Technology Development, July 20. <http://www.nsf.gov/od/cawmset/meetings/hearing-990720/wawulf/wawulf.htm>.

Xie, Y., and K. A. Shauman. 1998. Sex differences in research productivity: New evidence about an old puzzle. *American Sociological Review* 63:847–870.

Yanico, B. J., and S. I. Hardin. 1986. College students' self-estimated and actual knowledge of gender traditional and nontraditional occupations: A replication and extension. *Journal of Vocational Behavior* 28:229–240.

Young, C. 1999. *Ceasefire: Why Women and Men Must Join Forces to Achieve True Equality*. New York: Free Press.

Yount, K. R. 1991. Ladies, flirts, and tomboys: Strategies for managing sexual harassment in an underground coal mine. *Journal of Contemporary Ethnography* 19:396–422.

Zernike, K. 2000. Girls a distant 2nd in geography gap among U.S. pupils. *New York Times,* May 31.

Zhang, J., J. Quan, and P. Van Meerbergen. 1994. The effect of tax-transfer policies on fertility in Canada, 1921–88. *Journal of Human Resources* 29:181–201.

Ziller, R. C. 1957. Vocational choice and utility for risk. *Journal of Counseling Psychology* 4:61–64.

Zimmer, L. 1988. Tokenism and women in the workplace: The limits of gender-neutral theory. *Social Problems* 35:64–77.

Zucker, K. J., S. J. Bradley, G. Oliver, and J. Blake. 1996. Psychosexual development of women with congenital adrenal hyperplasia. *Hormones and Behavior* 30:300–318.

Zuckerman, H. 1991. The careers of men and women scientists: A review of current research. In H. Zuckerman, J. R. Cole, and J. T. Bruer (eds.), *The Outer Circle: Women in the Scientific Community*. New York: W. W. Norton.

Zuckerman, M. 1987. All parents are environmentalists until they have their second child. *Behavioral and Brain Sciences* 10:42–44.

Index

About the Author

Kingsley R. Browne is a professor at Wayne State University Law School. He has written extensively about biological sex differences and the glass ceiling, sexual harassment, and the military, as well as about statistical proof of discrimination and free speech. He is also author of *Divided Labours*. He lives in Ann Arbor, Michigan.